Fuels and Petrochemicals Division 2015

Core Programming Area at the 2015 AIChE Annual Meeting

Salt Lake City, Utah, USA
8-13 November 2015

ISBN: 978-1-5108-1862-0

Printed from e-media with permission by:

Curran Associates, Inc.
57 Morehouse Lane
Red Hook, NY 12571

Some format issues inherent in the e-media version may also appear in this print version.

Printed by Curran Associates, Inc. (2016)

For permission requests, please contact AIChE
at the address below.

AIChE
120 Wall Street, FL 23
New York, NY 10005-4020

Phone: (800) 242-4363
Fax: (203) 775-5177

www.aiche.org

Additional copies of this publication are available from:

Curran Associates, Inc.
57 Morehouse Lane
Red Hook, NY 12571 USA
Phone: 845-758-0400
Fax: 845-758-2634
Email: curran@proceedings.com
Web: www.proceedings.com

TABLE OF CONTENTS

404863 Production of Synthetic Fuel from Coal and Biomass

Abstract:

Ceramatec is the prime contractor on a project for the US Department of Energy to produce a ~2 BPD Fischer Tropsch pilot plant. The feedstock for the plant will be a mixture of coal and biomass with the objective of producing synthetic fuel with a carbon footprint that is below that of conventional petroleum based fuels. Ceramatec is currently operating a small, ¼ BPD Gas-to-Liquids pre-pilot plant at its Salt Lake City, Utah facility. The FT technology associated with this plant will be used as the basis for the CTL/BTL pilot plant. The paper will describe the currently operating GTL facility and results related to catalyst and FT testing. Further, the planned operation of the 2 BPD pilot facility will be described with the accompanying computer simulation of the pilot.

425457 Exploring the Synergies in Process Engineering Between Catalytic Fast Pyrolysis and Processes Based on Conventional Energy Feedstocks

Abstract:

Conventional energy processes based on petroleum feedstocks (e.g. Fluid Catalytic Cracking, Hydroprocessing, Hydrocracking and Delayed Coking) can play a vital role in facilitating understanding of the process engineering aspects for in *situ* and *ex situ* catalytic fast pyrolysis vapor upgrading. The higher oxygen content of biomass and its tendency towards coke and gaseous species formation serves as a major differentiating factor between conventional petroleum refining processes and biomass-based thermochemical processes like pyrolysis.

The development of 21 billion gallons of advanced biofuels annually by 2022 is mandated by the U.S. Energy Security and Independence Act of 2007. The potential of *in situ* and *ex situ* upgrading of fast pyrolysis vapors as one of the promising research options for thermochemical conversion of biomass to liquid transportation fuels has been explored in a recent NREL and PNNL thermochemical research pathway process design report (http://www.nrel.gov/docs/fy15osti/62455.pdf). Based on the understanding derived from process engineering literature on Fluid Catalytic Cracking, the possible scale-up implications of a significant coke content (~10% of biomass) observed in current experimental studies on catalytic fast pyrolysis vapor upgrading can be extrapolated. This presentation shall also reflect upon the differences noticed in conventional petroleum hydrodesulfurization and biomass hydrodeoxygenation reactions, from a process engineering standpoint.

428747 Catalytic Pyrolysis of Used Tyres for Clean Fuel

Abstract:

Catalytic pyrolysis of used tyres for clean fuel

Julius I. Osayi. [1], Sunny Iyuke[1], Diakanua N. Bavon[1], Jaco Van Der Walt[2] and Samuel E. Ogbeide[3]

[1]Nanotechnology Research Group, University of the Witwatersrand, Johannesburg, South Africa

- ojgrants@yahoo.com, Sunny.iyuke@wits.ac.za Diakanua.nkazi@wits.ac.za

[2]South Africa Nuclear Energy Corporation (NECSA), Pretoria, South Africa

Jaco.vanderwalt@necsa.co.za

[3]Department of Chemical Engineering, University of Benin, P.M.B 1154, Benin City, Nigeria

Samuelogbeide@gmail.com

One of the current challenges following the rapid demand for automotive product is the large quantities of used tyres that end up as non-biodegradable solid waste. This is a major environmental and economic problem [1-3]. Therefore in this paper, the catalytic pyrolysis of used-tyres in a fixed bed quartz tube reactor was performed. Zeolite served as the catalyst and the effect of the catalyst/tyre ratio on the yield and composition of the obtained pyrolytic oil (biocrude) was investigated. The biocrude, produced an at operating temperature of 600 ^{o}C and heating rate of 15 ^{o}C/Min was characterized by FT-IR, ^{1}H NMR and GC-MS followed by fractional distillation. The results of the boiling points showed that the liquid fractions are comparable to petroleum fuels and can serve as an alternative source of clean fuel [4-8] than the used tyre biocrude which has the set back of sulphur and tars.

Key words: Used tyres; Pyrolysis; Catalyst; Distillation and Fuels

Reference:

[1] Al-Salem S.M., Lettieri P., and Baeyens J (2009): *Journal of Hazardous materials* 172, 1690 - 1694

[2] Berrueco C., Esperanza E., Mastral F.J., Ceamanos J., and Garcia-Bacaicoa P (2005): pyrolysis of waste tyres in an atmospheric static-bed batch reactor: Analysis of the gases obtained. *J.Anal. Appl. Pyrolysis* 74, 245 – 253

[3] Islam M.R., Tushar M.S.H.K and Haniu H (2008). Production of Liquid fuels and chemicals from pyrolysis of Bangladeshi bicycle/rickshaw tire waste. *J.Anal. Appl. Pyrolysis* 82, 96-109

[4] Aydin H., and Ilkilic (2012). Optimization of fuel production from waste vehicle tires by pyrolysis and resembling to diesel fuel by various desulfurization methods. *Fuel,* 102, 605 – 612.

[5] Lopez G., Aguado R., Olazar M., Arabiourrutia M., and Bilbao J (2009): Kinetics of scrap tyre pyrolysis under vacuum conditions. *Waste management,* 29, 2649-2655

[6] Biofuels, Green Energy and Technology (2009): *Doi*: 10, 1007/978-1-84882-011-1-6

[7] Brammer J., and Bridgewater A.V (1999): A review of biomass drying technologies for thermal conversion. *Renewable and sustainable Energy Reviews*; 3(4): 243-289.

[8] Bhatia V.K., Mittal K.G., Mehrotra R.P., and Mehrotra M (1989): Hydrocarbon Fuels

from biomass. *Biomass* 4, 151-4

430722 Production of JP-8 Jet Fuel from Coal and Coal/Biomass Mixtures

Abstract:

An indirect liquefaction process employing advanced process intensification approaches is being developed to reduce costs and accelerate commercialization of a coal to liquid (CTL) process for converting coal and coal-biomass mixtures to JP-8 jet fuel. In this Air Force and DOE/NETL sponsored cooperative agreement, Southern Research is leading a team consisting of Precision Combustion, Inc. (PCI), IntraMicron, Inc., Southwest Research Institute, and Nexant, Inc., along with Chevron as selective Fischer-Tropsch (FT) catalyst supplier and the National Carbon Capture Center (NCCC) as the host site for coal or coal/biomass gasifier slip-stream testing. The process reduces the number of parasitic syngas cleaning steps and eliminates costly upgrading of wax produced in conventional indirect liquefaction processes, resulting in improved economics and lower lifecycle greenhouse gas (GHG) emissions for CTL-based jet fuel production. These improvements directly reduce capital costs of indirect CTL processes, resulting in significantly improved economics and cost competitiveness of CTL when compared to petroleum refining. Specific objectives of the project are 1) to improve a compact, pressurized, high temperature reformer developed by PCI to convert tar and light hydrocarbons to syngas, and deliver the required hydrogen (H_2) to carbon monoxide (CO) ratio for Fischer-Tropsch (FT) synthesis in the presence of H2S and other syngas contaminants; and 2) to demonstrate a selective and active second generation cobalt-zeolite hybrid Fischer-Tropsch (FT) catalyst from Chevron that predominantly produces liquid range C5-C20 hydrocarbons with high productivity (5X conventional FT catalysts) and jet fuel selectivity while eliminating the production and associated handling of solid C21+ wax. This work will present results from lab-scale development of H_2S tolerant reforming catalysts as well as from a long-term test of the hybrid FT jet-selective catalyst using a nominal 5 gallons per day skid-mounted fixed-bed FT reactor system installed at NCCC.

TEXAS A&M
UNIVERSITY at QATAR

Systematic Approach to Design Tailor Made Fuel Blends that Meets ASTM Standards

Saad Intikhab, Sawitree Kalakul, Hanif Choudhury, Rafiqul Gani and
Nimir Elbashir

CHEMICAL ENGINEERING PROGRAM
PETROLEUM ENGINEERING PROGRAM

TEXAS A&M
UNIVERSITY at QATAR

Introduction

Motivation of Work

- Designing surrogates for both diesel and gasoline
- A surrogate fuel is one that comprises of a small number of diverse pure compounds that is made in order to emulate certain characteristics of a target fuel that contains a large number of compounds [1]

Objective behind developing surrogate fuel:

- Simpler composition can facilitate insights into fuel composition and property effects
- Also has value as a time-invariant reference fuel for experimental studies
- Difficult to model conventional/GTL fuels as they are composed of hundreds to thousands of compounds. Our surrogate fuel can be used by other groups in their computational work like for simulation of combustion in internal combustion engines [1][2]

CHEMICAL ENGINEERING PROGRAM

TEXAS A&M
UNIVERSITY at QATAR

Preparation of Surrogates

Designing Fuel

- Computer Aided Model developed by Technical University of Denmark (DTU)
- Composition of Main Ingredient (MI) of diesel and gasoline along with additives were determined through the developed model

Blending to Prepare Surrogate Mixtures

- Blends were prepared representing diesel surrogate
- Blends of gasoline were developed with the help of industrial partners

Experimental Analysis According to ASTM Standards

- Diesel surrogate blends were tested for numerous properties viz. *distillation characteristics, density, viscosity, heat content etc.*
- Different properties viz. *vapor pressure, density, viscosity, heat content, compositional attributes etc.* were determined for model gasoline

CHEMICAL ENGINEERING PROGRAM

TEXAS A&M UNIVERSITY at QATAR

Diesel Surrogate Fuel Composition

- Two diesel surrogates were defined using computational model

Chemicals (MI Model 1)	Chemicals (MI Model 2)
Iso-octane	Iso-cetane
Cyclo-octane	Cyclo-octane
Tetralin	Tetralin
Tetradecane	Tetradecane
Dodecane	Dodecane

Table 1: Main Ingredient (MI) Diesel Composition

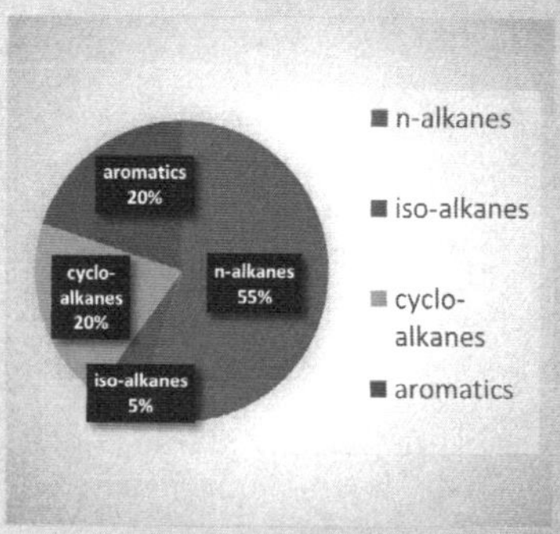

Figure 1: Hydrocarbon Groups (in vol%)

CHEMICAL ENGINEERING PROGRAM
PETROLEUM ENGINEERING PROGRAM

4

TEXAS A&M UNIVERSITY at QATAR

Test Methods and Equipment Used

Property	ASTM Method
Density	ASTM D4052
Composition by GC	Custom
Viscosity	ASTM D7042
Pour Point	ASTM D5949
Cloud Point	ASTM D5773
Flash Point	ASTM D93
Vapor Pressure	ASTM D6378
Heat content	ASTM D240
Distillation	ASTM D86
Cetane Index	ASTM D613

Table 2: Properties tested for surrogates

Density & Viscosity

Cloud Point & Pour Point

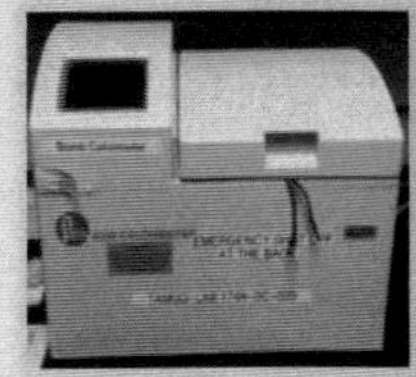
Heat Content

Flash Point

Distillation

Composition by GC

Comparison of MI with Conventional Diesel

Property	MI Model 1	MI Model 2	Conventional Diesel
Density (kg/m^3) at 15°C	813.1	817.4	829.5
Kinematic Viscosity (mm^2/s) at 40°C	1.5033	1.6117	3.1033
Heating Value (MJ/kg)	42.9607	43.2029	43.1739
True Vapor Pressure (kPa) at 37.8°C	1.3	0.5	0.2
Pour Point (°C)	-33	-33	-9
Cloud Point (°C)	-28.2	-30.3	-1.5
Flash Point (°C)	38.5	54.5	76.0
Cetane Index (Calculated)	42.89	42.85	58.16

Table 3: Fuel properties comparison of surrogate and conventional diesel

Gasoline Surrogate Fuel Composition

- Gasoline surrogates were defined using computational model [3]

Chemicals (MI)	Composition (wt%)
Butane	6.58
Heptane	12.60
Iso-octane	53.99
1-Pentene	3.63
Methylcyclopentane	8.47
Toluene	14.73

Table 4: Main Ingredient (MI) Gasoline Composition

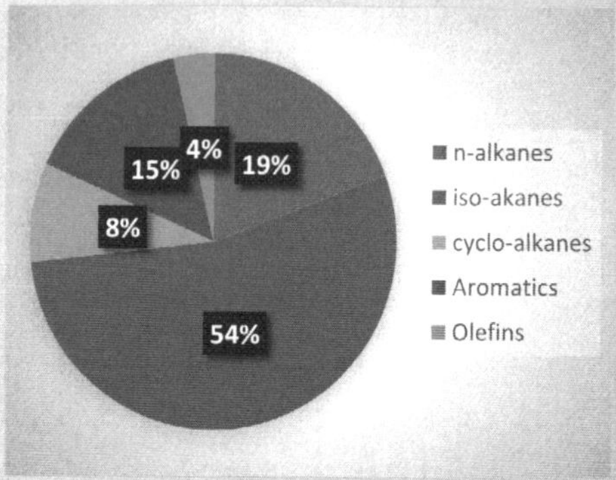

Figure 2: Hydrocarbon Groups (in wt%)

CHEMICAL ENGINEERING PROGRAM

TEXAS A&M UNIVERSITY at QATAR

Gasoline Surrogate Fuel Composition

- Additives for MI gasoline were also determined through computational model [3]

Blend Name	Blend Components
Blend 1	MI, Tetrahydrofuran, 2-Methyl Tetrahydrofuran
Blend 2	MI, Acetone, 2-Methyl Tetrahydrofuran
Blend 3	MI, Acetone, 2-Butanone
Blend 4	MI, 2-Butanone, 2-Methyl Tetrahydrofuran
Blend 5	MI, Ethanol, 2-Methyl Tetrahydrofuran

Table 5: Blends of MI gasoline and additives

CHEMICAL ENGINEERING PROGRAM

TEXAS A&M UNIVERSITY at QATAR

Sampling Methodology

- Sampling methodology developed in order to prevent loss of n-butane and volatile compounds like 1-Pentene

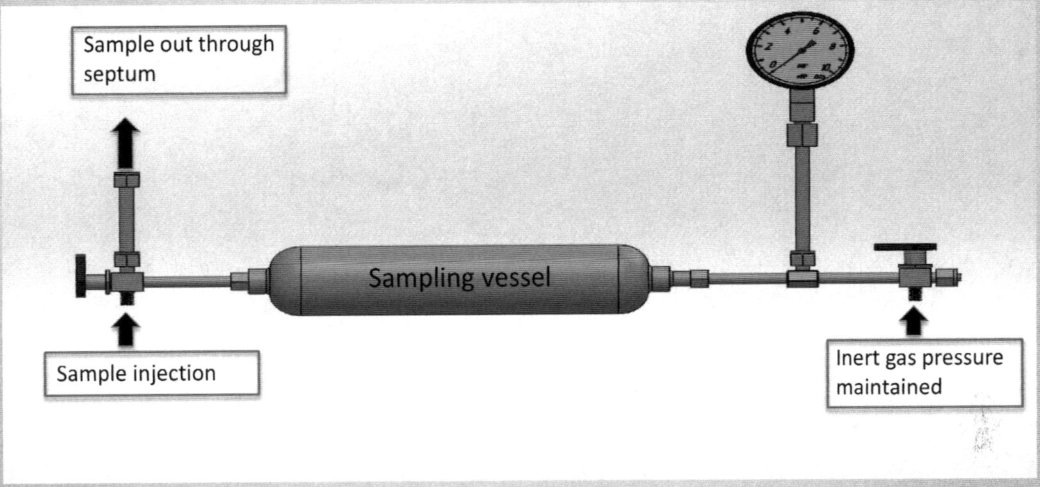

Figure 3: Sampling Device

DHA Analysis – Main Ingredient (MI)

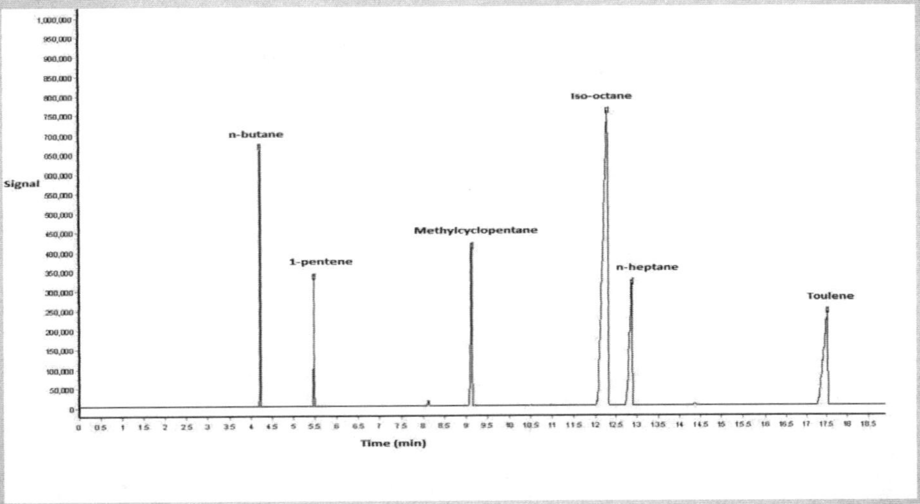

Figure 4: Gas Chromatogram of MI Gasoline

DHA Analysis – Blend 2

Component	Blended composition (wt%)	DHA composition (wt%)	Relative Standard Deviation (%)
N-butane	4.17	4.00	2.94
1-pentene	2.30	2.20	3.14
Methylcyclopentane	5.37	5.24	1.73
Iso-octane	34.21	34.07	0.29
N-heptane	7.98	8.06	0.71
Toluene	9.33	9.38	0.38
Acetone	13.77	12.97	4.23
2-Methyltetrahydrofuran	22.87	24.07	3.62

Table 6: Comparison of blended and DHA composition

Results of Model Gasoline

Sample	Density at 15°C $(\frac{g}{cm^3})$	Vapor Pressure at 37.8°C (kPa)	
Target Values	0.720 – 0.775	45 - 60	
	Experimental	Experimental	Model
Main Ingredient	0.7113	51.0	55.16
Blend 1	0.7596	46.2	46
Blend 2	0.7482	60.4	46
Blend 3	0.7333	58.9	49
Blend 4	0.7395	50.1	45
Blend 5	0.7357	55.1	45

Table 7: Fuel properties comparison with model results

Conclusion

- Study involves integration of computational modelling as well as experimental verification using advanced experimental techniques in developing surrogates for both diesel and gasoline
- Diesel surrogate MI Model 2 meets ASTM D975 for the properties measured
- The gasoline surrogates developed meets ASTM D4814 for the measured properties
- Sampling technique developed, effectively handles gas-liquid and volatile mixtures

CHEMICAL ENGINEERING PROGRAM | TEXAS A&M UNIVERSITY at QATAR

Thank you

Chemical Engineering Program and Petroleum Engineering Program, Texas A&M University at Qatar,

23874 Doha, Qatar

CHEMICAL ENGINEERING PROGRAM | TEXAS A&M UNIVERSITY at QATAR

References

1. Mueller CJ, Cannella WJ, Bruno TJ, Bunting B, Dettman HD, Franz JA, et al. Methodology for formulating diesel surrogate fuels with accurate compositional, ignition-quality, and volatility characteristics. Energy Fuels 2012;26:3284–303.
2. Ahmed A, Goteng G, Shankar VSB, Al-Quraishi K, Roberts WL, Sarathy SM. A Computational Methodology for Formulating Gasoline Surrogate Fuels with Accurate Physical and Chemical Kinetic Properties. Fuel 2015;143:290-300
3. Yunus NA, Gernaey KV, Woodley JM, Gani R. A Systematic Methodology for Design of Tailor-made Blended Products. Computers & Chemical Engineering 2014;66:201-13

15AIChE
Annual Meeting, Salt Lake City, UT

Solar Reactor Efficiency Analysis for Alternative Fuel Production via Ferrite Based Two- Step Thermochemical Splitting of Water and CO_2

(Principal Author)
Rahul Bhosale
Assistant Professor
Department of Chemical Engineering
College of Engineering
Qatar University, Doha, Qatar
rahul.bhosale@qu.edu.qa

(Other Authors)
Anand Kumar
Assistant Professor
Department of Chemical Engineering
College of Engineering
Qatar University, Doha, Qatar
akumar@qu.edu.qa

Fares A. AlMomani
Associate Professor
Department of Chemical Engineering
College of Engineering
Qatar University, Doha, Qatar
falmomani@qu.edu.qa

Dareen Dardor
Undergraduate Student
Department of Chemical Engineering
College of Engineering
Qatar University, Doha, Qatar
sg1000354@student.qu.edu.qa

Moustafa Hussein Ali
Undergraduate Student
Department of Chemical Engineering
College of Engineering
Qatar University, Doha, Qatar
ma1203561@student.qu.edu.qa

Shahd Gharbia
Department of Chemical Engineering
College of Engineering
Qatar University, Doha, Qatar
sg1000354@student.qu.edu.qa

Mehak Jilani
Department of Chemical Engineering
College of Engineering
Qatar University, Doha, Qatar
mj1104333@student.qu.edu.qa

Shiva Yousefi
Undergraduate Student
Department of Chemical Engineering
College of Engineering
Qatar University, Doha, Qatar
sy1106491@student.qu.edu.qa

Ahmed AlNouss
Teaching Assistant/Graduate Research Assistant
Department of Chemical Engineering
College of Engineering
Qatar University, Doha, Qatar
a.alnouss@qu.edu.qa

Mohammad Saad Anis
Undergraduate Student
Department of Chemical Engineering
College of Engineering
Qatar University, Doha, Qatar
ma1304725@student.qu.edu.qa

Mohammad Hamza Usmani
Undergraduate Student
Department of Chemical Engineering
College of Engineering
Qatar University, Doha, Qatar
mu1104953@student.qu.edu.qa

Manar Yousif Ali
Undergraduate Student
Department of Chemical Engineering
College of Engineering
Qatar University, Doha, Qatar
ma1105864@student.qu.edu.qa

Zineb Bouazza Bouabidi
Undergraduate Student
Department of Chemical Engineering
College of Engineering
Qatar University, Doha, Qatar
zb1103931@student.qu.edu.qa

Prepared for Presentation at
American Institute of Chemical Engineers
2015 Annual Meeting
Salt Lake City, Utah
November 8-13, 20

Solar Reactor Efficiency Analysis for Alternative Fuel Production via Ferrite Based Two- Step Thermochemical Splitting of Water and CO_2

Rahul Bhosale
Department of Chemical Engineering
College of Engineering
Qatar University, Doha, Qatar
rahul.bhosale@qu.edu.qa

Anand Kumar, Fares A. AlMomani, Dareen Dardor, Moustafa Hussein Ali, Shahd Gharbia, Mehak Jilani, Shiva Yousefi, Ahmed AlNouss, Mohammad Saad Anis, Mohammad Hamza Usmani, Manar Yousif Ali, Zineb Bouazza Bouabidi

Department of Chemical Engineering
College of Engineering
Qatar University, Doha, Qatar

Keywords: Ferrite Cycle, Water Splitting, CO_2 Splitting, Solar Syngas, Computational Thermodynamic Modeling

Abstract

In this paper computational thermodynamic modeling of solar syngas production via ferrite based (Fe_3O_4/FeO) thermochemical H_2O and/or CO_2 splitting cycle. This two-step cycle involves solar thermal reduction of Fe_3O_4 into FeO and O_2 and non-solar splitting of H_2O and CO_2 via exothermic oxidation of FeO into Fe_3O_4. Solar syngas produced will be used for the production of solar liquid transportation fuels by catalytic Fischer Tropsch process. This paper reports the solar reactor efficiency analysis for the iron oxide based solar syngas production process. Second-law thermodynamic analysis is applied to determine the cycle efficiencies attainable with and without heat recuperation. Also, the energy required for the inert Ar heating during the solar thermal reduction step was included during the efficiency calculation. Thermodynamic analysis has been performed and the analysis shows that solar syngas production via the Iron Oxide redox cycle is a promising approach for CO_2/H_2O conversion into alternative fuels.

1. Introduction

With the increase in the population of world and oil prices due to the depletion in the fossil fuel sources, production of solar syngas for the production of liquid transportation fuels is considered as one of the promising options. Solar radiation is an essentially inexhaustible energy source that delivers about 100,000 TW to the earth. To harvest the solar radiation and to convert it effectively into renewable fuels from H_2O and captured CO_2 provides a promising path for a future sustainable energy economy.

The present work considers the approaches for syngas production through solar fuels via the ferrite based redox cycle (Fe_3O_4/FeO). In order to evaluate the capability of the system for exporting work, the system was analyzed on the basis of the second law of thermodynamics and the reaction mechanism of ferrite taking into consideration that for high quality syngas. In previous investigations, various MO based redox systems were investigated for the solar thermochemical H_2O and/or CO_2 splitting such as ZnO/Zn, SnO_2/SnO, Fe_3O_4/FeO, mixed ferrites, and ceria based oxides [1 – 48].

In this investigation, thermodynamic analysis of the Fe_3O_4/FeO based solar thermochemical H_2O and CO_2 splitting process was investigated in two sections with the help of commercial thermodynamic HSC Chemistry software and databases. The second law thermodynamic analysis of this process was performed. The solar absorption efficiency of the solar reactor ($\eta_{absorption}$), solar energy input to the solar reactor (Q_{solar}), net energy absorbed in the solar reactor ($Q_{reactor-net}$), rediation heat losses from the solar reactor ($Q_{re-radiation}$), rate of heat rejected to the surrounding from the quench unit $\left(Q_{quench}\right)$, and from the FeO oxidizer ($Q_{FeO-Oxd}$) , and solar to fuel energy conversion efficiency $\left(\eta_{solar-to-fuel}\right)$ were determined by performing the thermodynamic modeling and are presented in this paper. A typical Fe_3O_4/FeO based two-step solar thermochemical H_2O and CO_2 splitting process is presented in Figure 1.

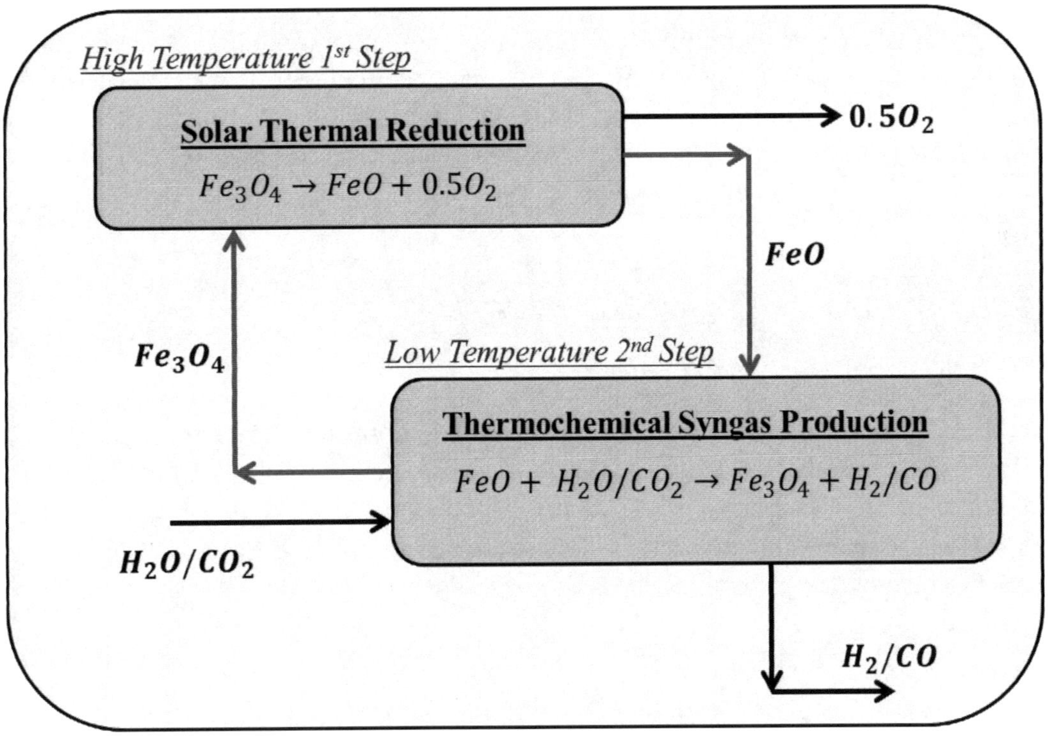

Figure 1. A typical Fe_3O_4/FeO based two-step solar thermochemical H_2O and CO_2 splitting process.

2. Solar Reactor Efficiency Analysis

The solar reactor efficiency analysis of the ferrite based solar syngas production via water and CO_2 splitting reaction was performed by using HSC Chemistry software and databases. The continuous Fe_3O_4/FeO based syngas production process is consists of a solar reactor, a quench unit, a FeO oxidizer, gas separator, and theoretical H_2/CO/O_2 ideal fuel cell. To perform the exergy analysis, following assumptions were made:

- The solar reactor is assumed to be a perfectly insulated blackbody absorber
- No convective/conductive heat losses
- Effective emissivity and absorptivity equal to unity
- All the products separate naturally without expending any work
- All reactions reach 100% completion
- Syngas production is carried out at atmospheric pressure and at steady state conditions
- Viscous losses and variation in the kinetic and potential energies are neglected
- Energy required for the separation of gaseous products are not accounted during the efficiency calculation
- Heat exchangers required for recovering the sensible latent heat are omitted from thermodynamic considerations

The exergy analysis was carried out by following the methodology and governing equations derived and used previously for other MO based solar thermochemical cycles. Thermodynamic properties were extracted from HSC software and databases and the analysis is normalized to the Fe_3O_4 molar flow rate (1 mol/sec) entering the solar reactor. Following reaction chemistry was followed for the solar reactor efficiency analysis.

$$Fe_3O_4 = 3FeO + 0.5O_2 \tag{1}$$

$$3FeO + 0.5H_2O + 0.5CO_2 = Fe_3O_4 + 0.5H_2 + 0.5CO \tag{2}$$

Absorption efficiency of ferrite based solar reactor

In previous investigations, it was reported that the complete solar thermal reduction of Fe_3O_4 into FeO is possible at around 2000K in presence of Ar = 100 mol/sec. The solar energy absorption efficiency ($\eta_{absorption}$) is defined as the net rate at which energy is being absorbed by the solar reactor divided by the solar power input through the aperture of the solar reactor. The $\eta_{absorption}$ can be calculated as:

$$\eta_{absorption} = \frac{Q_{reactor-net}}{Q_{solar}} = 1 - \left(\frac{\sigma T_H^4}{IC}\right) \tag{3}$$

Where,

σ = Stefan-Boltzmann constant,

I = normal beam insolation $1(kW/m^2)$,

C= solar flux concentration ratio of the solar concentrating system (1000 suns), $Q_{reactor-net}$= net energy absorbed in the solar reactor,

Q_{solar} = solar energy input.

For SnO_2/SnO based syngas production process, at T_H= 2000K, the $\eta_{absorption}$ was observed to be equal to 81.85%.

Net solar energy absorbed by the ferrite based solar reactor $(Q_{reactor-net})$

1 mole of Fe_3O_4 enters the solar reactor at T_R = 298K and then heated upto T_H = 2000K with the help of solar energy entering in the solar reactor from the solar concentrating system and undergoes 100% thermal reduction. The net solar energy absorbed by the solar reactor can be calculated as:

$$Q_{reactor-net} = Q_{Fe_3O_4-heating} + Q_{Fe_3O_4-reduction} + Q_{\text{Ar-heating}} \quad (4)$$

Thermodynamic calculations reveal that the 421.16 kW of heat is required for the heating of Fe_3O_4 from 298K upto 2000K. Furthermore, the amount of solar energy utilized for the thermal reduction of Fe_3O_4 at 2000K equals to 275.8 kW. Also 3537.77 kW is required to heat Ar from 298K upto 2000K. Therefore, the $Q_{reactor-net}$ for the solar reactor used for this process is equal to 4234.73 kW.

Solar energy input to the ferrite based solar reactor (Q_{solar})

With the help of calculated $\eta_{absorption}$ and $Q_{reactor-net}$, the amount of solar energy input required to operate the ferrite based solar reactor (Q_{solar}) dealing with the Fe_3O_4/FeO based solar thermochemical syngas production process can be determined by equation (5) and observed to be equal to 5173.77 kW.

$$Q_{solar} = \frac{Q_{reactor-net}}{\eta_{absorption}} \quad (5)$$

Radiation heat losses from the ferrite based solar reactor

Radiation heat losses from the solar reactor conducting the solar thermal reduction of Fe_3O_4 can be calculated by equation (6):

$$Q_{re-radiation} = Q_{Solar} - Q_{reactor-net} \quad (6)$$

For the solar reactor conducting the Fe_3O_4/FeO based solar thermochemical H_2O and CO_2 splitting process, the $Q_{re-radiation}$ equals to 939.04 kW.

Rate of heat rejected to the surrounding due to quenching of ferrite

After performing thermal reduction of Fe_3O_4 products exiting the solar reactor are rapidly cooled down from thermal reduction temperature (2000K) to ambient temperature (298K). The purpose of this quenching step is to avoid the recombination of the products. The rate of heat rejected by the quenching unit (Q_{quench}) during this step is calculated by the following equation (7) and observed to be 3918.16 kW.

$$Q_{quench} = -\dot{n}\Delta H|_{3FeO+0.5O_2(g)+\,100\text{Ar(g)}@2000K \rightarrow 3FeO+0.5O_2(g)+\,100\text{Ar(g)}@298K} \quad (7)$$

Rate of heat rejected to surrounding by the FeO Oxidizer

FeO from the quench unit is transferred to the FeO oxidizer and reacted with the combined stream of H_2O and CO_2 to produce syngas. It is noteworthy that the heat released from the exothermic H_2O and CO_2 splitting reaction is sufficient enough to overcome the sensible and latent heats and raise the temperature to the required splitting temperature. For this step, complete conversion of FeO to Fe_3O_4 is assumed and the rate of heat rejected to the surrounding from the FeO oxidizer is estimated to be 32.18 kW.

$$Q_{FeO-oxd} = -\dot{n}\Delta H|_{3FeO+0.5H_2O+0.5CO_2(g)\rightarrow Fe_3O_4+0.5H_2(g)+0.5CO(g)} \quad (8)$$

Rate of theoretical work performed and heat rejected by the Ideal fuel cell

The Fe_3O_4/FeO based syngas production cycle is completed with the addition of an ideal fuel cell operated in presence of H_2/CO/O_2. The fuel cell is added to the cycle to extract the maximum work of the net syngas produced by this process. The rates of theoretical work performed ($W_{FC-Ideal}$) and the heat rejected by the fuel cell ($Q_{FC-Ideal}$) can be calculated by using following equations (9) and (10). For This calculation the efficiency of the fuel cell (η_{FC}) is assumed to be 100%.

$$W_{FC-Ideal} = -(\eta_{FC}) \times \dot{n}\,\Delta G|_{0.5H_2(g)+0.5CO(g)+0.5O_2(g)\rightarrow 0.5H_2O+0.5CO_2(g)} \quad (9)$$

$$Q_{FC-Ideal} = -(T_R) \times \dot{n}\,\Delta S|_{0.5H_2(g)+0.5CO(g)+0.5O_2(g)\rightarrow 0.5H_2O+0.5CO_2(g)} \quad (10)$$

For the Fe_3O_4/FeO based solar thermochemical syngas production process, the $W_{FC-Ideal}$ and $Q_{FC-Ideal}$ are equal to 246.95 kW and 37.18 kW, respectively.

Cycle and solar-to-fuel energy conversion efficiency

Cycle efficiency of the solar thermochemical Fe_3O_4/FeO based syngas production cycle is defined as the ratio of theoretical work performed by the ideal fuel cell to the solar energy input:

$$\eta_{cycle} = \frac{W_{FC-Ideal}}{Q_{solar}} \quad (11)$$

Furthermore, the solar-to-fuel energy conversion efficiency of the solar thermochemical Fe_3O_4/FeO based syngas production cycle is defined as the ratio of higher heating value (HHV) of the H_2 and CO produced to the solar energy input:

$$\eta_{solar-to-fuel} = \frac{HHV_{0.5H_2+0.5CO}}{Q_{solar}} \quad (12)$$

As per the HSC computation thermodynamic modeling, η_{cycle} and $\eta_{solar-to-fuel}$ equal to 4.773% and 5.489%, respectively.

4. Conclusion

Two-step solar thermochemical cycle for syngas production via Fe_3O_4/FeO based H_2O and CO_2 splitting reactions have been thermodynamically examined. The solar reactor efficiency analysis of the solar reactor conducting the thermochemical Fe_3O_4/FeO based H_2O and CO_2 splitting process at 2000 K and at a solar concentration of 5000 suns indicate ${}_{\mathrm{absorption}} = 81.85\%$, $Q_{\mathrm{solar}} = 5173.77\ kW$, $Q_{\mathrm{reactor-net}} = 4234.73\ kW$, $Q_{\mathrm{re-radiation}} = 939.04\ kW$, $Q_{\mathrm{quench}} = 3916.16\ kW$, $Q_{\mathrm{FeO-oxd}} = 32.183\ kW$, and ${}_{\mathrm{solar-to-fuel}} = 5.489\%$, respectively.

5. Acknowledgments

The authors gratefully acknowledge the financial support provided by the Qatar University Internal Grants QUUG-CENG-CHE-13/14-4 and QUUG-CENG-CHE-14\15-10.

References

[1] Loutzenhiser, P., Steinfeld, A. Solar syngas production from CO_2 and H_2O in a two-Step Thermochemical Cycle via Zn/ZnO Redox Reactions: Thermodynamic cycle Analysis. International Journal of Hydrogen Energy, 36 (2011) 12141.

[2] Stamatiou, A., Loutzenhiser, P., Steinfeld, A. Solar Syngas Production from H_2O and CO_2 via two-Step Thermochemical Cycles based on Zn/ZnO and FeO/Fe_3O_4 Redox Reactions: Kinetic Analysis. Energy & Fuels, 24 (2010) 2716.

[3] Charvin, P., Abanades, S., Lemont, F., Flamant, G. Experimental study of $SnO_2/SnO/Sn$ thermochemical systems for solar production of hydrogen. AIChE Journal, 54 (2008) 2759.

[4] Abanades, S. CO_2 and H_2O reduction by solar thermochemical looping using SnO_2/SnO redox reactions: Thermogravimetric analysis. International Journal of Hydrogen Energy, 37 (2012) 8223.

[5] Chambon, M., Abanades, S., Flamant, G. Kinetic investigation of hydrogen generation from hydrolysis of SnO and Zn solar nanopowders. International Journal of Hydrogen Energy, 34 (2009) 5326.

[6] Vishnevetsky, I., Epstein, M., J. Tin as possible candidate for solar thermochemical redox process for hydrogen production. Journal of Solar Energy Engineering, 131 (2009) 021007-1.

[7] Abanades, S., Villafan-Vidales, H. CO_2 and H_2O conversion to solar fuels via two-step solar thermochemical looping using iron oxide redox pair. Chemical Engineering Journal 175 (2011) 368.

[8] Gokon, N., Murayama, H., Nagasaki, A., Kodama, T. Thermochemical two-step water splitting cycles by monoclinic ZrO_2 supported $NiFe_2O_4$ and Fe_3O_4 powders and ceramic foam. Solar Energy, 83 (2009) 527.

[9] Scheffe, J., Li, J., Weimer, A. A spinel ferrite/hercynite water-splitting redox cycle International Journal of Hydrogen Energy, 35 (2010) 3333.

[10] Roeb, M., Gathmann, N., Neises, M., Sattler, C., Pitz-Paal, R. Thermodynamic analysis of two-step solar water splitting with mixed iron oxides. International Journal of Energy Research, 33 (2009) 893.
[11] Lorentzou, S., Agrafiotis, A., Konstandopoulos, A., Aerosol spray pyrolysis synthesis of water-splitting ferrites for solar hydrogen production. Granular Matter, 10 (2008) 113.
[12] Abanades, S., Flamant, G. Thermochemical hydrogen production from a two-step solar-driven water-splitting cycle based on cerium oxides. Solar Energy, 80 (2006) 1611.
[13] Chueh, W. C., Falter, C., Abbott, M., Scipio, D., Furler, P., Haile, S., Steinfeld, A. High-Flux Solar-Driven Thermochemical Dissociation of CO_2 and H_2O using Nonstoichiometric Ceria. Science, 330 (2010) 1797.
[14] Meng, Q., Lee, C., Ishihara, T., Kaneko, H., Tamaura, Y. Reactivity of CeO2-based ceramics for solar hydrogen production via a two-step water-splitting cycle with concentrated solar energy. International Journal of Hydrogen Energy, 36 (2011) 13435.
[15] Le Gal, A., Abanades, S. Catalytic investigation of ceria-zirconia solid solutions for solar hydrogen production. International Journal of Hydrogen Energy, 36 (2011) 4739.
[16] Abanades, S., Gal, A. L. CO_2 Splitting by Thermo-Chemical Looping Based on $Zr_xCe_{1-x}O_2$ Oxygen Carriers for Synthetic Fuel Generation. Fuel, 102 (2012) 180.
[17] Bhosale RR, Shende RV, Puszynski JA, H2 generation from thermochemical water-splitting using sol-gel derived Ni-ferrite. Journal of Energy and Power Engineering 2010;4:27–38.
[18] Bhosale RR, Dardor D, Gharbia S, Folady J, Jilani M, Kumar A, van den Broke LJP, Lin F, Alexneit I, Thermochemical Conversion of CO_2 into Solar Fuels Using Ferrite Nanomaterials. In: Proceedings of 4th International Gas Processing Symposium 2015;141-185.
[19] Bhosale RR, Yousefi S, Folady J, Jilani M, Dardor D, Gharbia S, Kumar A, van den Broke LJP, Alexneit I, Shende R, Thermodynamic Analysis of Solar Fuel Production via Thermochemical H_2O and/or CO_2 Splitting Using Tin Oxide Based Redox Reactions. In: Proceedings of 4th International Gas Processing Symposium 2015;39 – 48.
[20] Bhosale RR, Jilani M, Dardor D, Gharbia S, Folady J, Kumar A, van den Broke LJP, Lin F, Alexneit I, Solar Fuel Production via Non-Stoichiometric $Ce_xZr_yHf_zO_{2-\delta}$ Based Two-Step Thermochemical Redox Cycle. In: Proceedings of 4th International Gas Processing Symposium 2015;117 – 124.
[21] Kumar A, Ashok A, Kuti YO, Bhosale RR, van den Broke LJP, Single Step Synthesis of Transition Metal Nanoparticles in Aqueous Phase for Catalytic Applications. In: Proceedings of 4th International Gas Processing Symposium 2015;69 – 80.
[22] Dowass NA, Alazewi NS, Shurair MS, Bhosale RR, van den Broke LJP, Kumar A, Modeling of reaction front movement in combustion synthesis for catalyst preparation. In: Proceedings of 4th International Gas Processing Symposium 2015;391 – 399.
[23] Dardor D, Bhosale RR, Gharbia S, Kumar A, AlMomani F, Solar Carbon Production via Thermochemical ZnO/Zn Carbon Dioxide Splitting Cycle. Journal of Emerging Trends in Engineering & Applied Sciences 2015;6:129 – 135.
[24] Bhosale RR, Kumar A, AlMomani F, Alexneit I, Propylene Oxide Assisted Sol-Gel Synthesis of Zinc Ferrite Nanoparticles for Solar Fuel Production. Ceramics International 2015 (in press).

[25] Dardor D, Bhosale RR, Gharbia S, Kumar A, AlMomani F, Solar Thermochemical Conversion of CO_2 into C via SnO_2/SnO Redox Cycle: Thermodynamic Study. International Journal of Engineering Research & Applications 2015;5:134 – 140.

[26] Bhosale RR, Kumar A, AlMomani F, Gharbia S, Dardor D, Ali M, Folady J, Yousefi S, Jilani M, Solar Thermochemical Reforming of Natural Gas via Metal Oxide Based Redox Reactions. In: Proceedings of 2015 AIChE Spring Meeting (ISBN: 978-0-8169-1089-2) 2015.

[27] Bhosale RR, Kumar A, AlMomani F, Gharbia S, Dardor D, Ali M, Folady J, Yousefi S, Jilani M, Al-Fakih M, A Novel Two-Step Metal Oxide – Metal Sulfate Water Splitting Cycle for the Production of Solar Hydrogen. In: Proceedings of 2015 AIChE Spring Meeting (ISBN: 978-0-8169-1089-2) 2015.

[28] Bhosale RR, Kumar A, AlMomani F, Dardor D, Gharbia S, Ali M, Folady J, Yousefi S, Jilani M, Abuarja D, Angre P, Solar Hydrogen Production via Thermochemical Metal Oxide – Metal Sulfate Water Splitting Cycle. In: Proceedings of 2015 AIChE Spring Meeting (ISBN: 978-0-8169-1089-2) 2015.

[29] Dardor D, Bhosale RR, Kumar A, AlMomani F, Gharbia S, Ali M, Folady J, Yousefi S, Jilani M, Computational Thermodynamic Analysis of Solar Fuel Production via Metal Oxide Based H_2O and CO_2 Splitting Thermochemical Cycles. In: Proceedings of 2015 AIChE Spring Meeting (ISBN: 978-0-8169-1089-2) 2015.

[30] Gharbia S, Bhosale RR, Kumar A, AlMomani F, Dardor D, Ali M, Folady J, Yousefi S, Jilani M, Al-Fakih M, Angre P, Thermochemical Conversion of H_2O and CO_2 into Solar Fuels via Metal Oxide Based Redox Reaction. In: Proceedings of 2015 AIChE Spring Meeting (ISBN: 978-0-8169-1089-2) 2015.

[31] Ashok A, Kumar A, Bhosale RR, Saleh MAH, van den Broeke LJP, Cellulose Assisted Combustion Synthesis of Porous Cu-Ni Nanopowders. RSC Advances 2015;5:28703 – 28712.

[32] Kumar A, Cross A, Manukyan K, Bhosale RR, van den Broeke LJP, Miller JT, Mukasyan AS, Wolf EE, Combustion Synthesis of Copper-Nickel Catalysts for Hydrogen Production from Ethanol. Chemical Engineering Journal 2015;278:46 – 54.

[33] Bhosale RR, Kumar A, van den Broeke LJP, Dardor D, Gharbia S, Jilani M, Folady J, Al-Fakih M, Tarsad M, Solar Hydrogen Production via Thermochemical Iron Oxide – Iron Sulfate Water Splitting Cycle. International Journal of Hydrogen Energy 2015;40:1639 – 1650.

[34] Bhosale RR, Alxneit I, Kumar A, van den Broeke LJP, Folady J, Dardor D, Gharbia S, Jilani M, Al-Fakih M, Elshrbajy NN, Tarsad M, Almkhyari FA, Solar Thermochemical Conversion of CO_2 into Synthetic Fuels via Ferrite Based Redox Reactions. In: Proceedings of 2014 AIChE Annual Meeting (ISBN: 978-0-8169-1086-1) 2014.

[35] Bhosale RR, Alxneit I, Kumar A, van den Broeke LJP, Jilani M, Folady J, Dardor D, Gharbia S, Dejan R, Al-Emadi N, Al-Massih F, Al-Yazedi S, Thermochemical Dissociation of CO_2 into Renewable Fuels via $Ce_xZr_yHf_zO_2$ Based Redox Reactions Using Concentrated Solar Energy. In: Proceedings of 2014 AIChE Annual Meeting (ISBN: 978-0-8169-1086-1) 2014.

[36] Kumar A, Ashok A, Kuty YO, Bhosale van den Broeke LJP, Combustion Synthesis of Copper-Nickel Catalysts for Hydrogen Production from Ethanol. In: Proceedings of 2014 AIChE Annual Meeting (ISBN: 978-0-8169-1086-1) 2014

[37] Bhosale RR, Alxneit I, van den Broeke LJP, Kumar A, Jilani M, Gharbia S, Folady J, Dardor D, Sol-Gel Synthesis of Nanocrystalline Ni-Ferrite and Co-Ferrite Redox Materials for Thermochemical Production of Solar Fuels. Materials Research Society Symposium Proceedings 2014;1675:mrss14-1675-rr06-10.

[38] Bhosale RR, Shende RV, Puszynski JA, Sol-Gel Derived $NiFe_2O_4$ Modified With ZrO_2 for Hydrogen Generation from Solar Thermochemical Water-Splitting Reaction. Materials Research Society Symposium Proceedings 2012;1387:mrsf11-1387-e09-07.

[39] Pasala X, Bhosale RR, Amar VS, Puszynski JA, Shende RV, Thermally Stabilized Ferrite Nanoparticles for Hydrogen Generation from Thermochemical Water-Splitting Reaction. In: Proceedings of 2012 AIChE Annual Meeting (ISBN 978-0-8169-1073-1) 2012.

[40] Bhosale RR, Pasala X, Yelakanti S, Puszynski JA, Shende RV, Solar Thermochemical Water-Splitting for H_2 Generation Using Sol-Gel Derived Ferrite Nanomaterials. In: Proceedings of NSTI – Nanotech Conference & Expo., (ISBN: 978-1-4665-6276-9) 2012;3:555.

[41] Bhosale RR, Shende RV, Puszynski JA, Synthesis of Nano-Sized Ferrite Materials for H2 Generation from High Temperature Water-Splitting Reaction. In: Proceedings of NSTI – Nanotech Conference & Expo., (ISBN: 978-1-4398-7142-3) 2011;1:237.

[42] Bhosale RR, Shende RV, Puszynski JA, H_2 Generation from Thermochemical Water-Splitting Using Sol-Gel Synthesized Ferrites ($M_xFe_yO_z$, M = Ni, Zn, Sn, Co, Mn, Ce). In: Proceedings of 2010 AIChE Annual Meeting (ISBN: 978-0-8169-1065-6) 2010.

[43] Bhosale RR, Shende RV, Puszynski JA, Sol-Gel Synthesis of Ferrite Foam Materials for H_2 Generation from Water-Splitting Reaction. In: Proceedings of NSTI – Nanotech Conference & Expo., (ISBN: 978-1-4398-3401-5) 2010;1:368.

[44] Bhosale RR, Shende RV, Opoku MK, Puszynski JA, H_2 Generation from Water-Splitting Using Novel Ferrite Foam Materials. In: Proceedings of 2009 AIChE Annual Meeting (ISBN: 978-0-8169-1058-6) 2009.

[45] Shende RV, Puszynski JA, Opoku MK, Bhosale RR, Synthesis of Novel Ferrite Foam Materials for Water-Splitting Application. In: Proceedings of NSTI – Nanotech Conference & Expo., (ISBN: 978-1-4398-1782-7) 2009;1:201.

[46] Bhosale RR, Shende RV, Puszynski JA, H2 generation from thermochemical water-splitting using sol-gel synthesized Zn/Sn/Mn-doped Ni-ferrite. International Review of Chemical Engineering 2010;2:852–62

[47] Bhosale RR, Khadka RP, Shende RV, Puszynski JA. H2 generation from two-step thermochemical water-splitting reaction using sol-gel derived SnxFeyOz. Journal of Renewable and Sustainable Energy 2011;3:063104-1–063104-16.

[48] Bhosale RR, Shende RV, Puszynski JA. Thermochemical water-splitting for H2 generation using sol-gel derived Mn-ferite in a packed bed reactor. International Journal of Hydrogen Energy 2012;37:2924–34.

433453 Aromatic Hydrocarbons for Jet Fuels from Lignin Fast Pyrolysis and Upgrading

Abstract:

Jet fuels have never been entirely substituted by biomass (viz. bio-jet fuel) due to the absence of aromatic hydrocarbons and cycloparaffins in biomass-derived oils. Currently, this problem can only be mitigated by blending bio-jet fuel with at least 50% traditional petroleum-based jet fuels. To produce 100% non-petroleum-based jet fuels, a suitable source of bio-based aromatic hydrocarbons and cycloparaffins is needed. Lignin is the only component in biomass rich in benzene-ring structures linked via ether bonds. Lignin is frequently underutilized due to its heterogeneity and resistance to chemical conversion; most of the lignin extracted from biomass is burned as low-value, stationary energy.

In this paper, we explore the opportunity of a four-step pathway to convert lignin into jet-fuel-range molecules aromatic hydrocarbons and cycloparaffins. Once produced, these molecules can be blended with bio-jet fuels to create 100% non-petroleum-based jet fuel that meets the technical specifications of Jet A. The production pathway consists of four steps: biomass pretreatment, lignin depolymerization, hydrodeoxygenation (HDO), and alkylation. During biomass pretreatment, lignin will be the final product rather than a waste. During HDO, the two-phase bio-crude oil derived from lignin depolymerization is used as feedstock. The bottom phase, dominated by heavier phenols as the main components from depolymerization, will be targeted. The mixture of lignin-derived bio-oils from HDO will serve as the feedstock for alkylation. In addition, the mutual influences of multi-components on desired product yields will be considered in HDO and alkylation of lignin-derived bio-oil. The review on each process is conducive to direct the pathway optimization to produce jet-fuel-range molecules from lignin in future research through a simplified and integrated pathway.

434035 Conversion of Furfural to Cyclopentanone over Ru/TiO_2

Abstract:

The conversion of biomass derived furanic species to useful compounds is a challenging task, especially in the presence of the water, carboxylic aicds, and phenolics that are commonly present in real streams. Under the severe conditions required for the stabilization of these oxygenated species, furanic compounds readily undergo humin and coke formation. In this contribution, we discuss the selective transformation of furanic species to the thermally stable cyclopentanone via a Piancatelli rearrangement. We discuss the role of active sites and co-adsorbed species on Ru/TiO_2 for this transformation in the vapor phase. The role of partially hydrogenated intermediates, surface bound water, as well as defects on the TiO_2 support are discussed. Model compound studies are presented with accompanying reaction kinetics. Results are also presented with real biomass torrefaction vapors, where we report the rare circumstance selectivity enhancement to cyclopentanone by co-adsorption of various oxygenates.

417297 Kinetic Study of Hydrodeoxygenation of Palmitic Acid in a Microreactor

Abstract:

Kinetic study of hydrodeoxygenation of Palmitic Acid in a Microreactor

Lin Zhou and Adeniyi Lawal

New Jersey Center for Microchemical Systems, Dept. of Chemical Engineering and Materials Science, Stevens Institute of Technology,

1 Castle Point on Hudson,

Hoboken, NJ 07030

Microalgae, on account of its high acylglycerides content and faster growth rate than terrestrial oil crops, are favored as a feedstock for green diesel production via hydrodeoxygenation (HDO). Performance studies on HDO of Nannochloropsis salina (N.S.), the workhorse of the industry, were conducted and reported previously.

Here we extend this previous work by studying the kinetics of palmitic acid, selected as the model compound for N.S. oil due to its high content in N.S., in a microreactor using 1% Pt/¦Ã-Al_2O_3. Mass and heat transfer limitations were firstly examined to make sure the rate data was obtained in the intrinsic kinetic regime. Kinetic experiments were conducted in a differential mode, in which the conversion of reactants was kept below 5% to assure the reactor was essentially gradientless. Both Power law and Langmuir-Hinshelwood type rate equations were fitted to the experimental data. Finally, the goodness of fit of the proposed kinetic models was determined by examining the statistical parameters.

434986 Catalytic Hydrothermal Liquefaction of Pinewood, Switchgrass and Wastepaper for H2, Biocrude and Bio-Oil Production

Abstract:

The authors report (a) catalytic hydrothermal liquefaction (HTL) of different biomass substrates (pinewood, switchgrass and wastepaper) at 200-275 °C in presence of a homogeneous nitrate salt catalysts: $Ni(NO_3)_2$, $Ca(NO_3)_2$, $Co(NO_3)_2$ and $Fe(NO_3)_3$ for the production of H_2, biocrude and bio-oil, (b) detailed investigation of biocrude and bio-oil generated from a biomass and (c) study of the possible reaction pathways for production of bio-oil with their energy content. To generate H_2 and biocrude, a slurry of biomass was prepared with different biomass to solvent (B:S) ratios of 1:10, 1:30, and 1:75 and loaded in the SS316 PARR reactor and heated to different temperatures. While the reaction was in progress, gas samples were withdrawn at regular time intervals and analyzed using GC equipped with thermal conductivity detector and Chrompack GC column. The product gas analysis revealed the presence of mainly H_2, CH_4, CO and CO_2. The processed slurry was centrifuged to recover biocrude and residue. The biocrude was further analyzed by TOC analyzer and HPLC. Among different catalysts used, $Ni(NO_3)_2$ yielded highest mol% H_2 at 275 °C and a maximum biocrude at 250 °C. The biocrude was found to contain sugars, C_1-C_3 carboxylic acids such as lactic, propionic, acetic and formic, HMF/furfural and other oxygenated compounds, which were analyzed by HPLC and GC–MS. The biocrude obtained was further processed to recover light bio-oil (LBO), heavy bio-oil (HBO) and residue which were found to have higher heating value (HHV)'s compared with original biomass. The results obtained on H_2, biocrude/bio-oil composition obtained during HTL of biomass substrates and possible reaction pathways will be presented in detail.

417175 Low Temperature Electrochemical Methods for Upgrading Bio-Oils

Abstract:

Bio-oil is a complex mixture of hundreds of organic and inorganic compounds. The reactive species in bio-oil complicate storage, transportation, and downstream processing because of secondary reactions. As such, these oils require upgrading to be used in the conventional transportation fuel infrastructure. The current design case for upgrading of pyrolysis oils uses a complex three-stage process, where the reactors are operated at high temperatures and up to 2000 psi of hydrogen. We have utilized electrochemical methods to upgrade bio-oils under very mild conditions without the need for externally-supplied hydrogen. Here we have used an electrolysis system to hydrogenate reactive functionalities to more stable compounds using protons and water as the hydrogen source. Electrochemical hydrogen generation is a common competing reaction. We have studied novel cell structures and electrode materials to improve the current efficiency for electrochemical reduction of organic species. Promising results have been observed where significant hydrogenation has been found. A description of work conducted and discussion of results will be presented.

429518 Cold Flow Study of a Segregated Zoning Burner

Abstract:

There are large deposits of oil shale and inaccessible coal in the US and around the world. Shale oil extraction is usually performed by ex-situ processing by mining the oil shale and treating it in processing facilities. New in-situ technologies can benefit from using gas-fired burners for supplying heat locally to facilitate extraction of the oil via oil walls. The purpose of this work is to develop a system for in-situ heating of oil shale and other kerogen-rich deposits. A new flameless segregated-zone burner has been proposed and is being evaluated by looking at mixing of fuel and oxidant over extended distances. A combination of engineering calculations and computational fluid dynamic (CFD) simulations are being used to evaluate the efficacy of different design options. A set of cold-flow mixing experiments are also being used to validate the design calculations. The results of these design efforts will be presented, along with an assessment of technical, economical and operational concerns for the proposed burner.

430723 Effect of Nanoporous Materials on Saturation Pressure in Shale Reservoirs

Abstract:

There has been a rapid growth in research for thermodynamic properties of shale reservoirs. The common pore size of shale reservoirs is approximately 1-20 nm and the interactions between the pore wall and the fluid within the pore may have a significant effect on the localized fluid properties and phase behavior in these confined spaces. The focus of this work is the effect of porous materials on the saturation pressure of a hydrocarbon mixture of methane and decane. Synthesized SBA-15 was utilized as the nanoporous material, which can represent the porous system in shale reservoirs.

The phase diagram of the binary mixture of methane and decane has been modeled with the Windprop module from the CMG reservoir simulator. The experimentally measured saturation pressure without nanoporous material matches well with the simulated results. However, the measured saturation pressure changes when the mixture is within the nanoporous material. This implies that the saturation pressure of hydrocarbon mixtures in shale reservoirs are different than those found in conventional reservoirs.

Evaluation of Cyclic Gas Injection EOR Performance on Shale Core Samples Using X-Ray CT Scanner

Yang Yu and James J. Sheng, Texas Tech University, Lubbock, TX;
Fersheed Mody, Apache Corporation, Houston, TX

Abstract

Production from tight and shale oil reservoirs will be more global in the coming decades. In order to efficiently produce unconventional reservoirs in a huff-n-puff injection mode, it is necessary to have a deeper understanding of the oil recovery mechanism. In this study, such potential is investigated by using CT imaging technique. It helps us to observe the process of oil production from the core samples and evaluate the recovery performance based on CT images.

Two groups of experiments were conducted, and two Wolfcamp core samples were used for each group. One group were saturated with mineral oil, the other one were saturated with crude oil. All samples were experienced 10 cycles of huff-n-puff N2 injection to recovery oil. The only variable parameter in each group was the soaking period that were 1 hour and 3 hours, respectively. In addition to determining recovery factor by measuring the core weight loss for each cycle, the core was imaged using X-ray CT to visualize the phase flow and fluid distribution inside the core. This study compared the RF results from shale core samples saturated with mineral oil and crude oil by cyclic gas injection EOR method. The work also introduced CT imaging during the experiment to assist us to understand the oil displacement process.

This study presents experimental results of cumulative recovery factor of core samples after each cycle. Samples were imaged at various recovery process of huff-n-puff. The results show that recovery factor of samples saturated with mineral oil under the soaking period of 3 hours and 1 hour are 32.9% and 24.1%, respectively. The recovery factor (RF) of samples saturated with crude oil under 3 hours and 1 hour soaking period are 23.7% and 27.9%. The obtained CT images, together with the analysis of CT number at various locations of the sample during the experiment reveal the density and composition of substance inside the plug and reflect the oil recovery. Results show that more mineral oil is produced under 3 hours soaking time compare with 1 hour soaking time with other operation parameters being the same, while the results are not the same for the case of cores saturated with crude oil. The difference in properties of core samples is believed to cause such consequence. So the RF is sensitive to rock and oil properties.

Introduction

With the tight reservoirs rapid development and production, it is necessary to maintain reservoir pressure and optimize tight oil recovery. For such naturally fractured with ultra-low permeability plays, pressure maintenance may only be achieved by the injection of gas rather than water.[1] Recently, researchers pay more attention to apply cyclic gas injection (CGI) to shale oil or shale liquid-rich reservoirs, the potential was evaluated by laboratory tests and simulation studies. Sheng[2] comprehensively reviewed and compared the EOR options in shale reservoirs. He concluded that gas injection has a better performance than water injection to recovery shale oil, and huff-n-puff gas injection is more effective than gas flooding.

The ideal choice of injection gas should be methane that could avoid to contaminate the reservoir hydrocarbon fluids; or carbon dioxide, which is available to achieve miscibility at a lower pressure than other gases. However, methane is in high demand for industrial and home uses, and carbon dioxide is not always available in required volumes. Air is easy to access and cost-effective, but we do not suggest inject it into reservoirs containing sour hydrocarbons, causing oxidizing reaction and facilities corrosion.[1] Therefore, N2 is an economical agent for oil recovery by gas displacement.

The earlier published experimental works on CGI[3-6] indicated that the shale oil recovery efficiency is sensitive to lots of operational parameters, such as injection pressure, soaking time, pressure depletion rate, and production time. For the lab-scale study, even though the recovery factor can be calculated by the means of mass balance, measuring the weight loss for each huff-n-puff cycle, it is still unclear about the oil production process during the cycle. How many oil recovered during the huff stage and what percentage of the production contributed by the puff stage? In this study, computerized tomography (CT) is applied to find the answer. CT is a non-destructive imaging technique that utilizes X-ray technology and mathematical reconstruction algorithms to view cross-sectional slices of an object[7]. As a useful tool, CT scanner have been widely used in petroleum engineering applications for determinations of in-situ saturation[8] and porosity[9], investigations of coreflood experiments[10] and evaluation of EOR processes[11, 12]. For this study, while performing huff-n-puff tests, core samples were scanned under the designed experiment conditions that allow us to observe the change of fluids composition in the pore areas.

The objective of this study is to understand oil-producing process of huff-n-puff EOR technique with the aided of X-ray CT imaging, and evaluate recovery efficiency of cyclic N2 injection under immiscible conditions on shale core samples.

Experimental Work

Materials

Wolfcamp outcrop core plugs were used as the shale core samples. They were cut in a direction parallel to the bedding plane with the dimension of 1.5 inch in diameter and 2 inch in length. Nitrogen was used as gas source to perform huff-n-puff tests.

The liquids for core saturation process were mineral oil Soltrol® 130 and crude oil from the field. Properties of the oil samples at the temperature of 20°C are given in Table 1.

Table 1. Oil Sample Properties

Oil Type	Density (g/cm^3)	Viscosity (cp)	Surface Tension (dynes/cm)	CT Number (Hounsfield)
Mineral oil	0.75	1.4	23	-255
Crude oil	0.85	8.5	25	-184

Setup and Procedures

All core samples were cleaned, dried, and weighted. Prior to perform CGI, cores were saturated with oil under the constant pressure of 1,000 psi for 24 hours. Then release the pressure, take the core out from vessel, and remove the excess liquid from the surface of the core sample. To minimize error, each sample was placed in the lab environment for 10 hours to stabilize the weight and equalize the internal and external pressure of core sample, then measure the weight gain. After the saturation process, cores were ready for conducting huff-n-puff tests. The weight loss of each sample was measured and recorded at the end of each cycle.

The apparatus set-up diagram and detailed experimental procedures of mineral oil saturation and huff-n-puff process were described in our previous work[6]. To avoid pump erosion or damage caused by crude oil, an accumulator was used to transfer energy in crude oil saturation process as shown in **Fig.1**.

Four series of tests were performed 10-cycle immiscible N2 huff-n-puff process under the constant operating pressure of 1,000 psi and the room temperature of 22°C. **Table 2** shows the test conditions. Two of core samples were saturated with Soltrol 130. For comparison, the other two were saturated with crude oil. Refer to the conclusion from our previous study[6] that recovery factor increases with the increase of pressure depletion, the blowout pressure depletion mode was applied in the puff stage (pressure depletion time was 0.05 hour). 6 hours was applied as production time to allow the core weight become stabilized.

Table 2. Test Conditions

Test No.	Core No.	Saturated Fluid	Soaking Time, hours
1	W#1	Soltrol 130	1
2	W#2	Soltrol 130	3
3	W#3	Crude oil	1
4	W#4	Crude oil	3

CT Scanning Process

A CT scanner was used to monitor the tests of 3 and 4 (**Fig. 2**). Instead of placing core in the high pressure vessel, an X-ray core holder was used without rubber sleeves inside so that the annular space between core plug and core holder can be simulated as natural fractures. The core holder was placed on the CT scan platform, inlet was connected with the N2 source, and outlet was plugged. The position of core sample was remained the same place throughout the experiment.

The CT scanner used in this study was CereTom^{TM} NL3000 manufactured by NeuroLogica Corp. Considering that core length was 2 inch (50.8mm), the coverage of scan was set as 50mm. To observe homogeneity of core sample, the slice thickness was set as the minimum value of 1.25mm, so 40 images were obtained after each scan.

For test #3, the scan interval was 10mins during the soaking period (huff stage), then perform scanning at the production period (puff stage) of 10mins and 10hrs, respectively, for total 10 cycles of huff-n-puff. The first cycle of test #4 was scanned per 20mins during the soaking period, and at the production period of 10mins and 60mins. Both samples were scanned at the end of each cycle to determine recovery factor using CT number.

Each substance and combination of substance, phases and systems is associated to a CT number. The CT number is a measure of the X-ray attenuation due to the sample placed in the X-ray beam[11]. During the test, the change in CT number reflects the change of substance composition in the pore area of core sample. In other words, it reflects the in-situ oil/gas saturation. Therefore, we will monitor and analyze the average CT number of core sample during the huff-n-puff process.

Core sample porosity can be calculated by using the CT number of the sample at the conditions of dry (0% saturating fluid) and saturated sample (100% saturated fluid), according to the following equation:

$$\varphi = \frac{CT_{sat} - CT_{dry}}{CT_{oil} - CT_{air}} \quad (1)$$

The oil phase saturation can be expressed as:

$$S_o = \frac{CT_{exp} - CT_{dry}}{CT_{sat} - CT_{dry}} \tag{2}$$

Where CT_{sat} is the image of the core sample 100% saturated with oil phase, CT_{dry} is the image of the cleaned and dried core sample, CT_{oil} is the CT number of the oil sample that is used for saturation, CT_{air} is the CT number of pure air. CT_{exp} is any CT number of the core sample obtained during the huff-n-puff process.

Results and Discussion

In this study, four tests of N2 huff-n-puff process under immiscible conditions were performed on mineral oil-saturated and crude oil-saturated core samples. Tests 3 and 4 were monitored by CT imaging throughout the recovery process. The porosity of core samples were determined, shale oil recovery factor were reported, and recovery performance of all tests and oil-producing process were discussed here.

Determination of Porosity

For all core samples, porosity was evaluated in two ways: saturation method and CT number method. Porosity is the ratio of pore volume to the bulk volume of the sample, and the pore volume is equal to the weight difference of dry and oil-saturated core over the density of saturated fluid. The results are shown in **Table 3**. CT imaging allows us to determine the porosity of each scanned slice. **Fig. 3** shows porosity heterogeneity. We found that the porosity measured from CT number is slightly higher than the value calculated from saturation method. W#3 has the highest porosity of 10.28%. Porosities of W#1 and #2 are very close.

Table 3. Porosity Determined from Two Measurement Methods

Core No.	Oil Saturation Method	CT Number Method
W#1	9.26%	9.54%
W#2	9.31%	9.60%
W#3	10.28%	11.38%
W#4	8.14%	8.66%

Oil-producing Process

To have a better understanding of immiscible CGI recovery mechanism, tests 3 and 4 were scanned by CT. **Fig. 4** shows the average CT number of W#3 during the huff-n-puff process for each cycle. **Fig. 5** displays the CT number of various slices of W#4 in the first cycle. In **Fig. 4**, for each single cycle, during the soaking period (huff stage), CT number initially presents a slight increase, then become stable or fluctuate in a narrow range; in the following production period (puff stage), CT number decreases apparently. **Fig. 5** gives the same information from the changes in CT number of various slices and whole core plug. CT number directly reflects the density of scanned substance. If it increases, it means higher density material entered the pore area of the core; if it decreases, it means the total density of substances in the pore area decreased.

At the beginning of huff stage, core was in the atmospheric pressure condition, surrounded by air; during the huff stage, core was surrounded by 1,000 psi N2; when puff stage began, the compressed N2 in the pore space of core rapidly spread out, differential pressure was the energy that pushes the oil out of core. The reason of CT number increased at the early period of huff stage was that high pressure N2 begin flow into the core. CT number decreased since puff stage represented that oil came out of the pore space. From the changes in CT number during each huff-n-puff process, we can infer that the major oil production stage was in the puff stage, and there was almost no oil recovered in the huff stage.

Recovery Performance

The recovery performance of huff-n-puff is evaluated by recovery factor (RF) based on material balance. It is the ratio of the produced oil to the original oil in the core. **Table 4** summarizes the results of cumulative RF obtained at the end of each recovery cycle for each one of the tests performed. The results also plotted in **Fig. 6**. For tests 1 and 2, both core samples, with similar properties of porosity, were re-saturated with mineral oil. After 10 cycles of N2 huff-n-puff process, the RF of W#1 and W#2 are 24% and 33%, respectively. It shows that longer soaking time is contribute to better recovery performance as demonstrated by previous studies[4,6].

However, W#3 under 1-hour soaking time recovered 28% oil and W#4 under 3-hour soaking time only recovered 24% oil. The RF of tests 3 and 4 were also determined by CT number as shown in **Fig. 7**. Although RF measured from CT is higher than that calculated from weight difference, **Fig. 7** shows similar recovery tendency and same found that W#3 produced more oil than W#4 as presented in **Fig. 6**. This is contrary to our initial expectation. Besides experimental errors, the main possible reason that causes the opposite results is core properties such as porosity and permeability. Even though permeability was not determined in this study, W#3 has about 2% higher porosity than W#4. Larger porosity brings more storage space and fluid channel for fluids in the pore area of core. The effect of core properties is superior to the effect of soaking time on the oil recovery. Comparing the RF results of W#2 (33%) and W#4 (24%), it can be estimated that mineral oil is easier to be recovered than crude oil under the same test conditions, because mineral oil has much lower viscosity than crude oil.

Table 4. Cumulative Shale Oil Recovery Factor for Each Test

Test No.	cycle 1	cycle 2	cycle 3	cycle 4	cycle 5	cycle 6	cycle 7	cycle 8	cycle 9	cycle 10
#1	5.41%	8.54%	10.92%	13.46%	15.41%	17.23%	18.94%	20.79%	22.34%	24.11%
#2	5.89%	10.94%	14.95%	18.29%	21.84%	24.72%	26.73%	29.01%	31.02%	32.94%
#3	7.64%	13.55%	17.07%	19.24%	21.18%	22.72%	24.50%	26.04%	27.00%	27.85%
#4	6.56%	10.94%	14.52%	16.33%	17.85%	19.25%	20.61%	21.90%	22.99%	23.71%

Conclusions

- Based on the analysis of weight loss and CT number, experimental results show that RF increased with the number of cycle performed. The incremental recovered oil decreased with the increase of gas injection cycles.
- The main oil-producing process occurred since the beginning of puff stage, no obvious oil production was shown during the huff stage.
- The effect of core properties on recovery performance is significantly. Sample in higher porosity tend to produce more oil with a higher recovery factor.
- Mineral oil that with lower viscosity is easier to be produced than crude oil.

Acknowledges

The work presented in this paper is supported by the U.S. Department of Energy under Award Number DE-FE0024311.

Reference

1. Mungan, N. (2000), "Enhanced Oil Recovery with High Pressure Nitrogen Injection," paper SPE 62547 presented at the 2000 SPE/AAPG Western Regional Meeting held in Long Beach, California, 19-23.
2. Sheng, J.J. (2015), "Enhanced Oil Recovery in Shale Reservoirs by Gas Injection," *J. Natural Gas Science and Eng.* 22 (1), 252-259.
3. Asghari, K. and Torabi, F. (2007), "Laboratory Experimental Results of Huff 'n' Puff CO2 Flooding in a Fractured Core System," paper SPE 110577 presented at the SPE Annual Technical Conference and Exhibition held in Anaheim, California, USA, 11-14 November.
4. Gamadi, T.D., Sheng, J.J., and Soliman, M.Y. (2003), "An Experimental Study of Cyclic Gas Injection to Improve Shale Oil Recovery," paper SPE 166334 presented at the SPE Annual Technical Conference and Exhibition held in New Orleans, Louisiana, USA, 30 September-2 October.
5. Wang, Z., et al. (2013), "Optimizing Cyclic CO2 Injection for Low- permeability Oil Reservoirs through Experimental Study," paper SPE 167193 presented at the SPE Unconventional Resources Conference held in Calgary, Alberta, Canada, 5-7 November.
6. Yu, Y. and Sheng, J.J. (2015), "An Experimental Investigation of the Effect of Pressure Depletion Rate on Oil Recovery from Shale Cores by Cyclic N_2 Injection" paper URTeC 2144010 presented at the Unconventional Resources Technology Conference held in San Antonio, Texas, USA, 20-22 July.
7. Siddiqui, S. and Khamees, A.A. (2004), "Dual-Energy CT-Scanning Applications in Rock Characterization," paper SPE 90520 presented at the SPE Annual Technical Conference and Exhibition held in Houston, Texas, USA, 26-29 September.
8. MacAllister, D.J., et al. (1993), "Application of X-Ray CT Scanning to Determine Gas/Water Relative Permeabilities," *SPE Formation Evaluation*, 8(3), 184-188.
9. Siddiqui, S., et al. (2005), "Techniques for Extracting Reliable Density and Porosity Data from Cuttings," paper SPE 96918 presented at the SPE Annual Technical Conference and Exhibition held in Dallas, Texas, USA, 9-12 October.
10. London, M., et al. (2014), "Waterflooding Experiments with X-ray CT Imaging," paper SPE 170147 presented at the SPE Heavy Oil Conference held in Alberta, Canada, 10-12, June.
11. Kovscek, A.R., Tang, G.Q. and Vega, B. (2008), "Experimental Investigation of Oil Recovery from Siliceous Shale by CO2 Injection," paper SPE 115679 presented at the 2008 SPE Annual Technical Conference and Exhibition held in Denver, Colorado, USA, 21-24 September.
12. Tovar, F.D., et al. (2014), "Experimental Investigation of Enhanced Recovery in Unconventional Liquid Reservoirs using CO2: A look Ahead to the Future of Unconventional EOR," paper SPE 169022 presented at the SPE Unconventional Resources Conference held in The Woodlands, Texas, USA, 1-3 April.

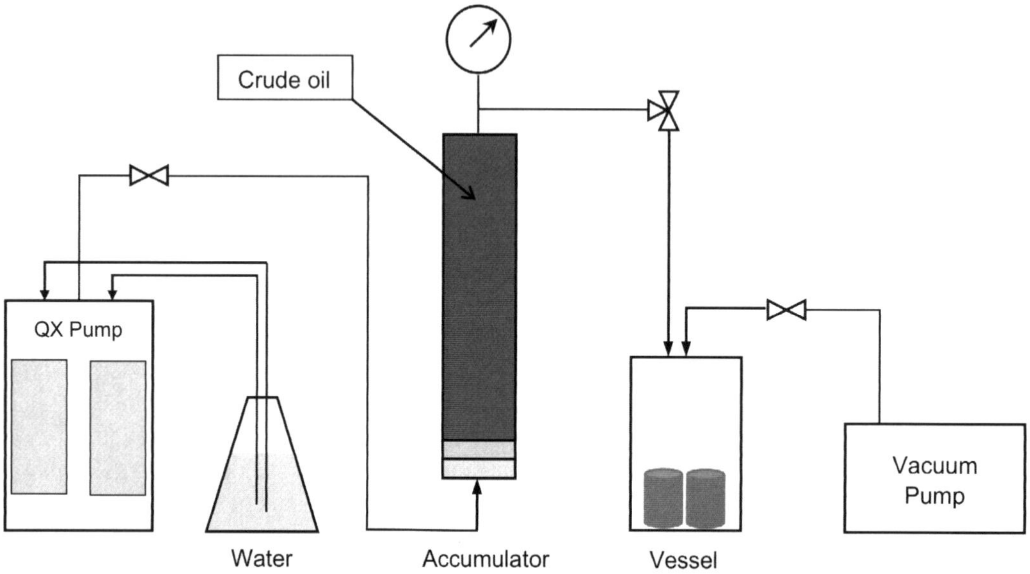

Figure 1. Experimental apparatus setup for crude oil saturation

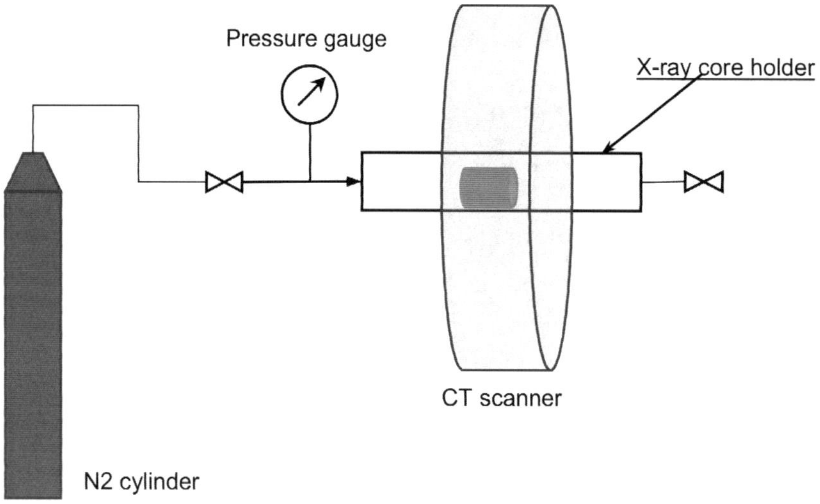

Figure 2. Diagram of experimental setup for huff-n-puff test

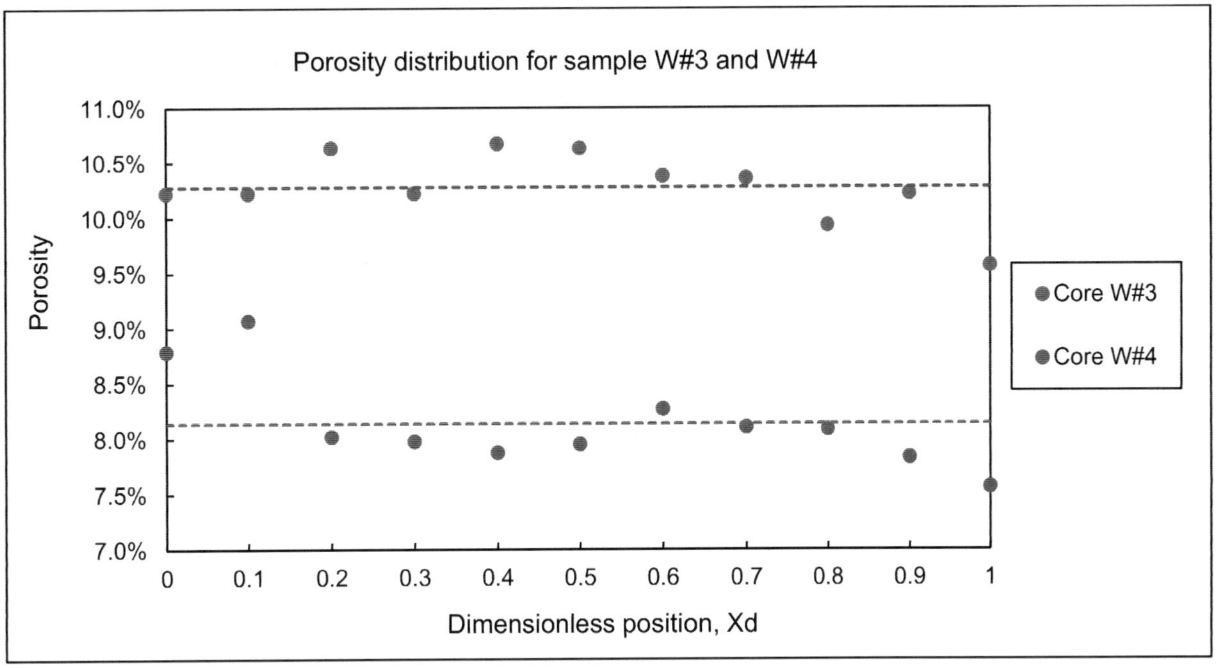

Figure 3. Average cross-sectional porosity versus dimensionless length for core sample 3 and 4

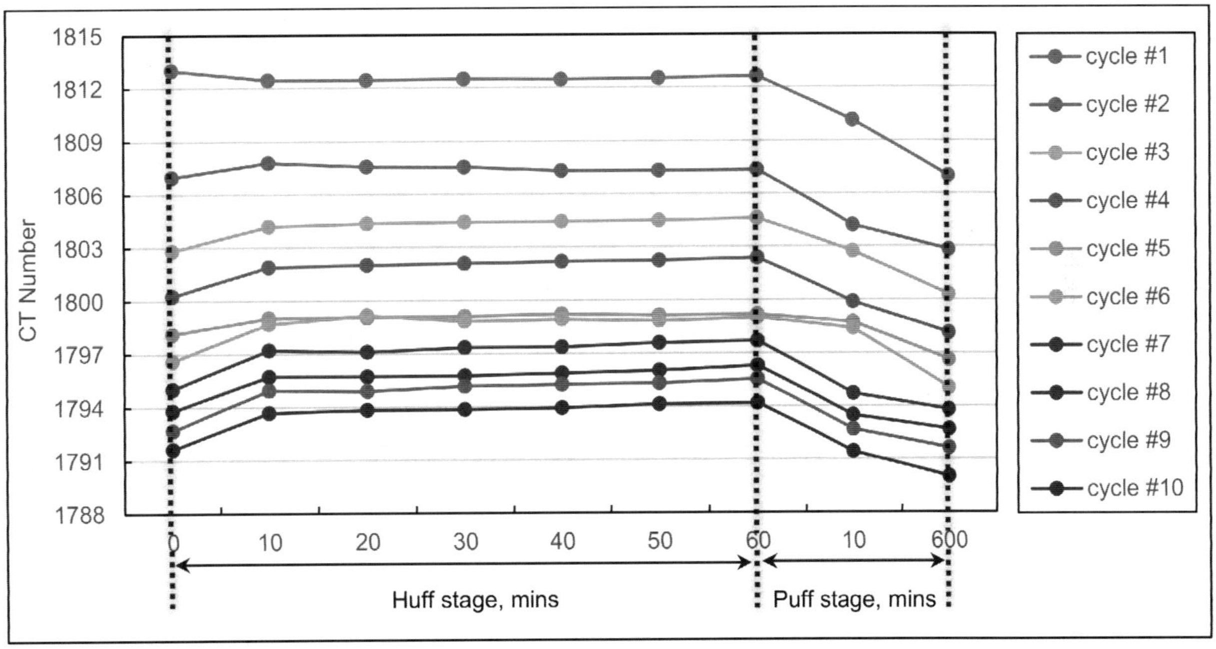

Figure 4. The changes of CT number throughout 10 cycles of huff-n-puff process for core W#3

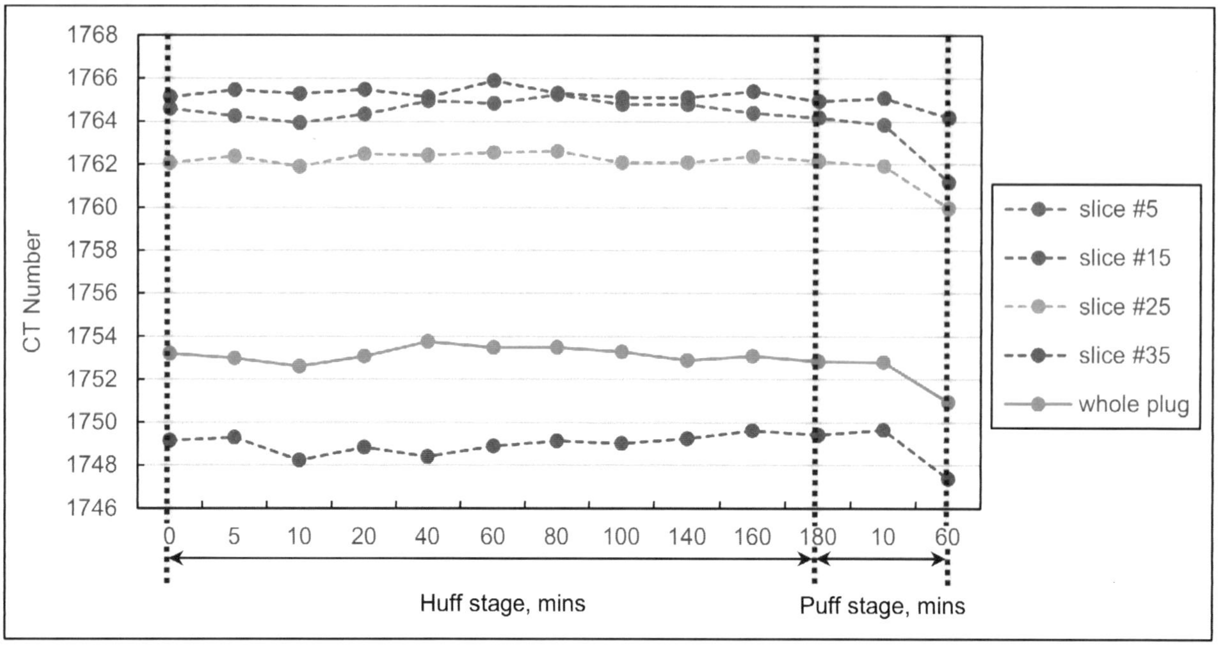

Figure 5. The changes of CT number in the 1st cycle of huff-n-puff for various locations of core W#4

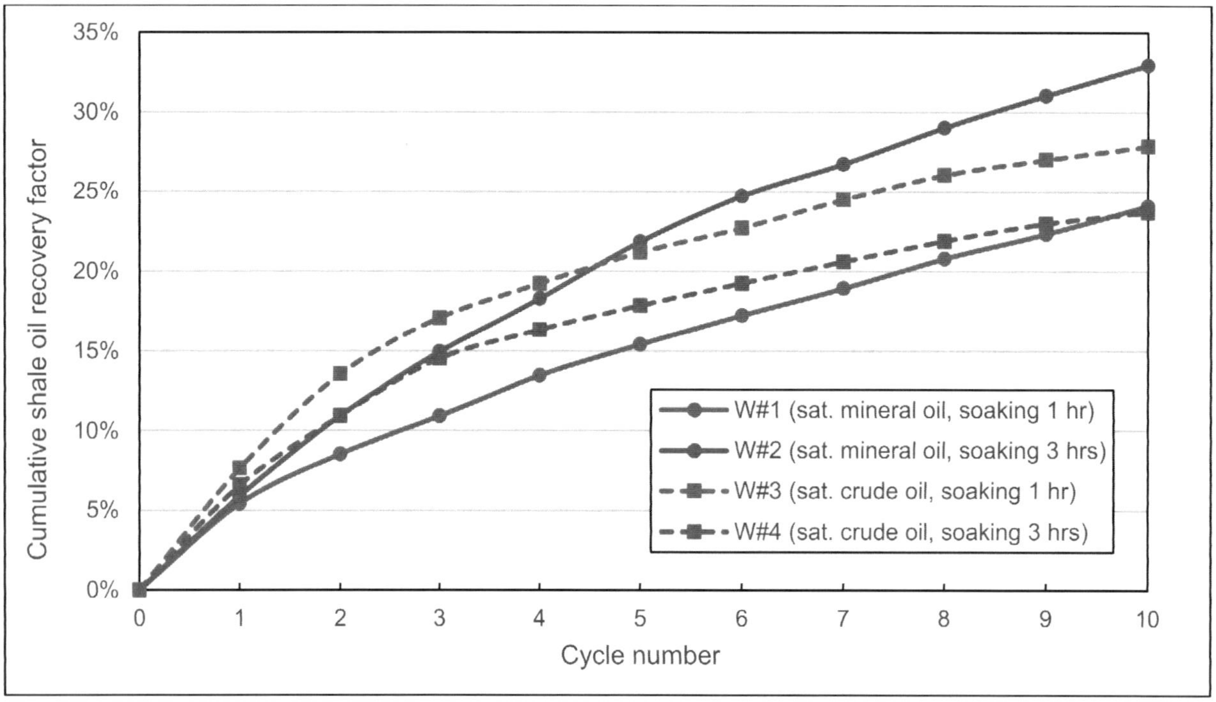

Figure 6. Cumulative oil recovery by N2 cyclic injection for all tests

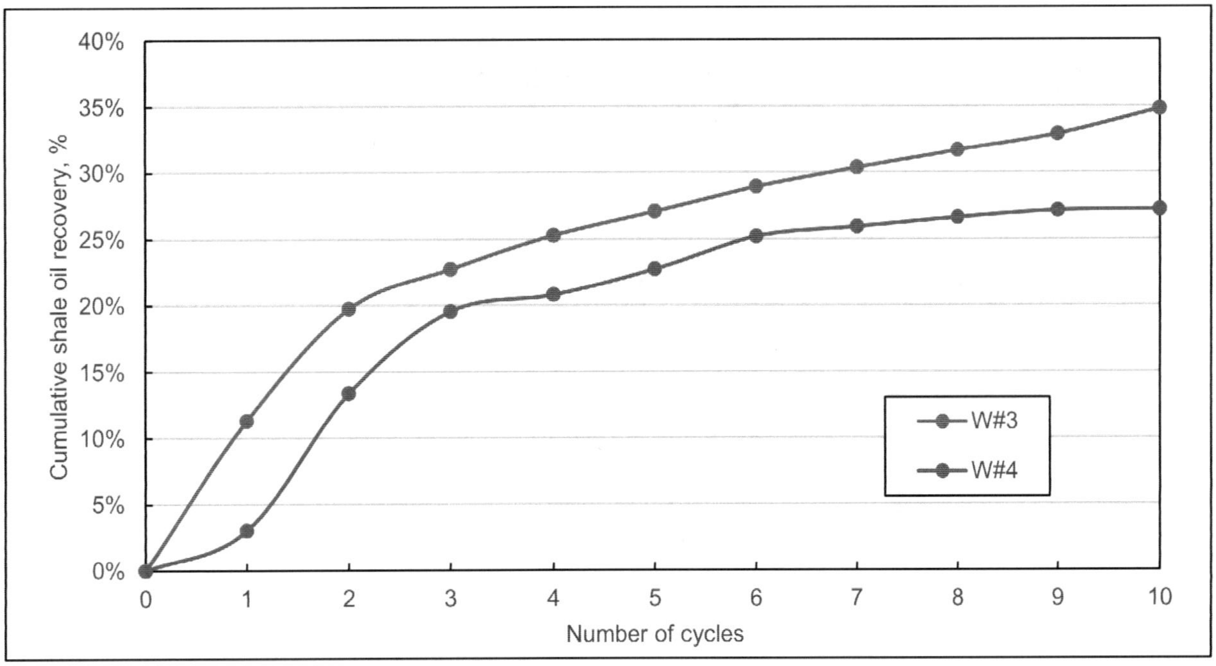

Figure 7. Evaluation of huff-n-puff recovery performance by CT scanner for test 3 and 4

Experimental and Numerical Upscale Study of Cyclic Methane Injection to Enhance Shale Oil Recovery

Lei Li, James J. Sheng, Marshall Watson, Texas Tech University,
Fersheed Mody, Apache Corporation

Abstract

Cyclic gas injection is considered as an effective and quick responding recovery process to enhance the shale oil recovery compared with water or gas flooding. And many finished lab experiments using Carbon dioxide and Nitrogen cyclic injection with core samples have proved its effectiveness. The impacts of oil saturation time, soaking time, depletion time, injection pressure, and cycle numbers have been tested in our previous work. However, the influence of core size on oil recovery is a very important parameter when upscale the core size to field level. Also, methane is a more economic and convenient gas source when considering the injection cost compared with carbon dioxide. In order to get the more accurate and reliable results, cyclic methane injection were used and the cores from Eagle Ford with different diameters (1 inch, 2 inches, 2.5 inches, 3 inches, 3.5 inches and 4 inches) were saturated with crude oil from Wolfcamp in this experiment. A numerical model was built using the typical reservoir properties from Eagle Ford to simulate cyclic methane injection. Sensitivity analysis, history matching and size upscale are conducted. The experiment results show that oil recovery factor decreases with the incremental of core sizes from 1 to 4 inches. Simulation results show that the most important parameter is injection pressure followed by matrix permeability, porosity and methane diffusivity. When upscale the core size to field level, the oil recovery decreased.

Introduction

Using a cyclic gas injection method to enhance oil recovery from conventional oil reservoir has been studied by numerous researches through laboratory studies, as well as field tests. In recent years, this technology has been applied to the production of unconventional reservoir. Its high potential in the shale oil reservoir development has been proved. Gamadi et al. (2013, 2014) conducted the nitrogen and CO_2 cyclic injection experiments with core plugs of Barnett, Marcos and Eagle Ford, and he claimed that injecting CO_2 at miscible pressure has great impact on the recovery factors compared to the injection at immiscible condition. He also tested the influence of soaking period on ultimate recovery factor and found the longer the shut-in period the more the ultimate recovery achieved[1,2]. Yu and Sheng (2015) analyzed the effect of pressure depletion time on shale oil recovery. They found that oil recovery increases as the pressure depletion time increases[3]. Wang et al. (2013) tested the effect of injection cycles number on oil recovery by CO_2 huff-n-puff experiments. They observed that the first cycle contributes nearly or above half of the total oil production of the operation and additional oil from subsequent cycles is significantly decreased[4]. Wan et al. (2013) conducted the simulation modeling of cyclic gas injection in shale oil reservoir and indicated that total oil recovery can be increased up to 29% by cyclic gas injection, compared with the 6.5% recovery from the primary depletion[5]. Yu et al. (2014) built a numerical fracture model to simulate CO_2 huff-n-puff in Bakken tight oil reservoirs and did the sensitivity analysis. Their results showed that the most important parameter is CO_2 injection rate, followed by CO_2 injection time, number of cycle, CO_2 diffusivity. The other parameters such as fracture conductivity, CO_2 soaking time, permeability and fracture half-length are less sensitive based on the range investigated in this study[6, 7].

Numerous papers have been published on cyclic gas injection, but our current knowledge of the field-scale performance of these processes is very limited. Reservoir simulation is required to properly scale up the process from laboratory to field conditions and to understand and interpret reservoir data. Previous work on upscale focused on conventional reservoir study. Mattax and Kyte (1962) proposed an upscale equation of oil recovery vs. dimensionless time for spontaneous imbibitions under the condition of same core sample shapes and fluid viscosity ratios[8]. Ma and Zhang (1995) improved this equation by giving a new definition of dimensionless time and they claimed this improved equation fitted with different porous media, core dimensionless, boundary conditions and oil, water viscosity[9]. Li and Horne (2002) generated another dimensionless time with almost all the parameters considered based on theoretical analysis of the fluid-flow mechanisms that govern spontaneous imbibition[10]. Mojdeh et al. (2009) upscaled the oil recovery results of the static imbibition experiment using a reference time based on gravity[11]. Pujol and Boberg (1972) discussed the scaling accuracy of laboratory steam flooding models and found that unscaled capillary pressures have negligible effect on oil recovery factor for high viscosity oil but will result in a higher prediction of oil recovery for the medium-viscosity oils[12]. Leung and Sanjay (2011) analyzed the uncertainty caused by scaleup of reservoir attributes and flow response in heterogeneous reservoirs[13]. Thomas et al. (1997) derived scaling criteria for micellar flooding using Dimensional Analysis and Inspectional Analysis. Their derivations were based on three-phase, six components flow in a porous medium[14]. Modeling these production process at large scales requires proper scaleup of petrophysical properties.

In this paper, we conducted the experiments using a new kind of gas Methane instead of Nitrogen or CO_2. The oil used to saturate the core is crude oil, not the mineral oil, which is similar to the field production. In the numerical study, a 9×9×6 gridblocks Cartesian model was built and sensitivity analysis was conducted. Then the model was validated with oil recovery data from our laboratory experiments. At last, the model was upscaled to field-scale and the cyclic gas injection for the field-scale model was simulated.

Experimental Section

In this paper, seven shale oil cores from Eager Fold with the same length of 2 inches, but different diameters with 1, 1.5, 2, 2.5, 3, 3.5 and 4 inches were used in our experiment. The injection gas is methane with purity more than 99%. All the experiment were performed at the temperature of 95°F in the oven. The experiment can be divided into two sections including the oil saturation section and cyclic gas injection section. According that traditional saturation method for conventional reservoir cores is not suitable to be used in these shale cores, a special saturation method applied in our previous experiments which is the same as Yu and Sheng[2] (2015) was used in this study.

Experimental procedures of core saturation (shown in Fig. 1):

1. Core samples are named and dry in the oven for one day.

2. Turn off the oven and leave the sample in the oven to the lab temperature, then weigh and record the sample as W1.
3. Place the core samples in the stainless steel vessel shown in Fig. 1, connect the vessel to a vacuum pump and vacuumed for three days.
4. Stop the vacuum pump and delivery water using the Quizix QX pump to push the crude oil in the accumulator, make sure core was completely soaked in crude oil and the saturation pressure in the vessel reach the requirement of 2000 psi.
5. After saturated for 2 days or a longer time, relief the soaking pressure and keep the cores inside of the vessel for one day to make sure the internal and external pressure of the cores reach an equilibrium condition.
6. Saturation process is finished. Weigh and record the saturated sample as W2.

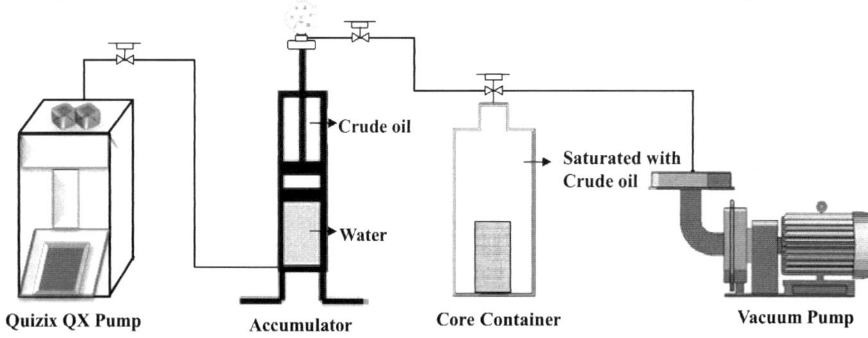

Fig. 1. Schematic diagram of the experimental equipment for oil saturation.

Experimental Procedures of Cyclic methane injection (shown in Fig. 2):

1. After oil saturation, put the core sample into the core holder in the oven with temperature of 95°F. Inject methane to the core holder. When the pressure gauge installed on the top of the container shows 2,000 psi, it means gas injection completed and soaking period starts.
2. Maintain the internal pressure of 2,000 psi for one day. This certain period of soaking time allows gas to dissolve into the oil.
3. After soak period finished, release the internal pressure to the atmosphere pressure of 14.7psi. Weigh and record the core sample again as W3.
4. Repeat 6 cycles and weigh the core after each cycle.

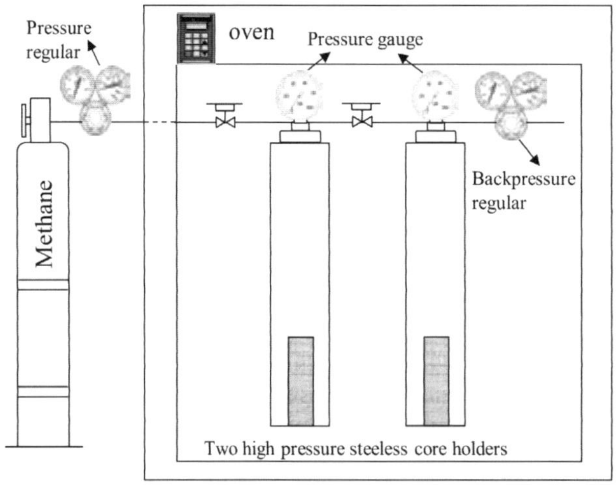

Fig. 2. Schematic diagram of the experimental equipment for cyclic gas injection.

Experimental Results

Table 1 includes the recovery factors of cyclic methane injection for shale cores at different core sizes with same length of 2 inches and different diameters of 1, 1.5, 2, 2.5, 3, 3.5 and 4 inches. Fig. 3 shows the incremental oil in different injection cycles for 1 inch core sample. We can see that the first cycle contributes to the most oil recovery and then the incremental oil recovery in one cycle decreases as the number of injection cycles increases. The incremental oil in the fifth cycle is only 4.67% which is much less than that of 12.63% in the first cycle. The experiments of other core sizes also reveal the same phenomenon. This indicates that the oil close to the core surface is easy to be produced than the deep oil inside the core. Fig. 4 displays the oil recovery in the first cycle for different diameter sizes of the core. All the experiments were conducted under the same operating situation (injection pressure 2000psi). The oil recovery for 1 inch core sample is 12.63% while 8.62% for 4 inches core sample in the first cycle. The incremental oil in one cycle declines with the increases of core diameters which show that the core with bigger diameter has lower oil recovery.

Table 1 Oil recovery data in gas injection cycles with different core diameters.

Core No.	Diameters, inches	Cycle 1	Cycle 2	Cycle 3	Cycle 4	Cycle 5
1	1	12.63%	20.84%	29.11%	35.41%	40.07%
2	1.5	11.26%	19.73%	27.34%	33.58%	39.17%
3	2	10.53%	18.58%	25.90%	32.86%	38.34%
4	2.5	10.13%	17.46%	24.87%	31.66%	37.18%
5	3	9.77%	16.31%	24.23%	30.92%	36.08%
6	3.5	9.34%	15.52%	23.14%	28.97%	33.79%
7	4	8.62%	14.46%	21.15%	27.69%	32.31%

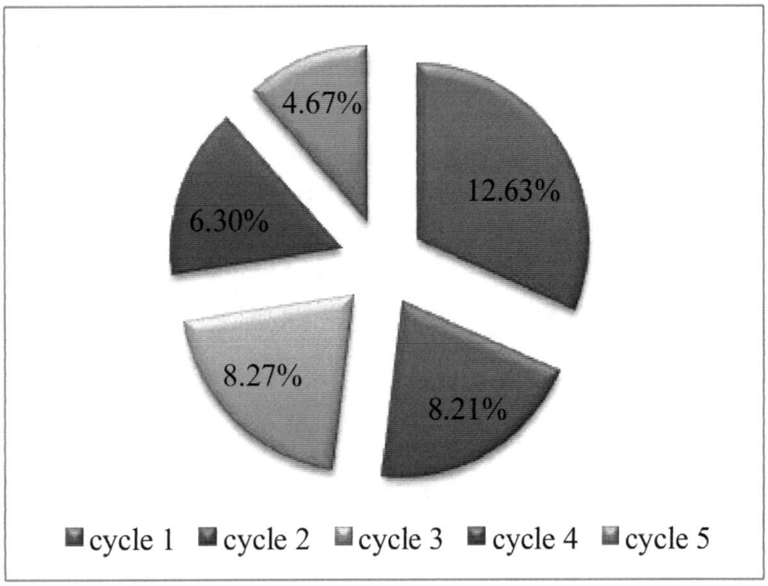

Fig. 3. Incremental oil in each injection cycle during cyclic methane injection for 1 inch core sample

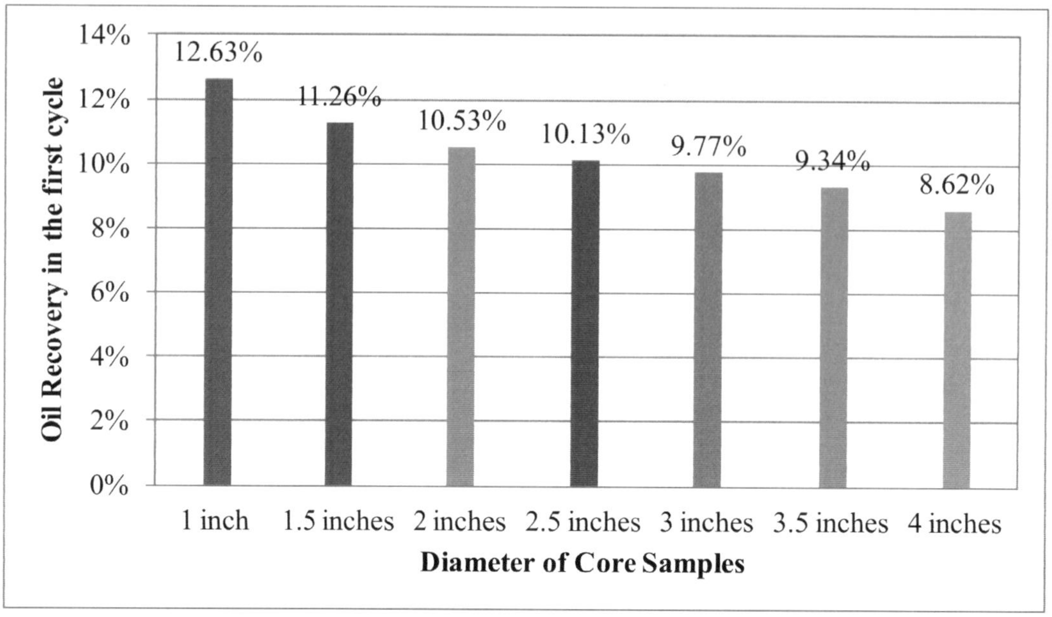

Fig. 4. R.F. data of different cores sizes for the first injection cycles under the conditions of 2000psi with crude oil and methane injection.

Numerical Simulation

A light crude oil from the Wolfcamp formation in Apache's Lin field was used in saturating the cores for the cyclic methane injection experiments. Anumerical model was built using CMG software to simulate cyclic methane injection shows in Fig. 5. For simplicity, the model is considered as homogeneous. The parameters of the model show in Table 2.

Table 2 Parameters used for this model

Parameter	Value	Unit
Matrix Permeability	0.0005	md
Fracture Permeability	1000	md
Porosity	6	%
Total compressibility	1×10^{-6}	psi^{-1}
Methane Injection Pressure	2000	psi
Injection time	0.01	day
Soaking time	1	day
Production time	0.15	day
Temperature	95	℉
Oil viscosity at laboratory	3.7	cp

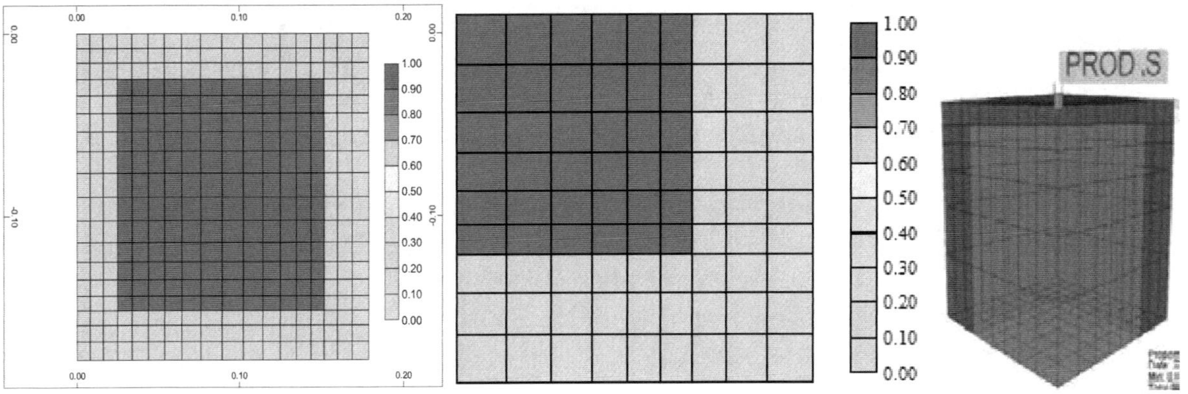

a. Gridblocks of 18*18*12 b. Gridblocks of 9*9*6 c. 3D Gridblocks of 9*9*6

Fig. 5. Gridblocks of Methane injection model

Fig. 5 a shows the original model with gridblocks of 18×18×12. It's a Cartesian model containing 2 wells, 1 producer and 1 injector to simulate the cyclic gas injection process in the experiment. Eighteen grid blocks were used in the X, Y direction and twelve grid blocks in Z direction. The matrix section S1 covers grid blocks from 4 to 15 in XY plane and 2 to 11 in Z direction. Other parts are considered as fracture section. In order to best represent the large pressure drops, gas and oil saturation changes near the matrix-fracture interface, logarithmic grids were used in this model with smaller grid blocks close to the surface and bigger grid blocks far away from the contact surface. The constant ratio (log(R1/R2))for adjacent grid blocks is 0.046. It will take about 3 hours to finish the simulation. To reduce the running time, a 1/8 model was used to replace the original total model as Fig.5 b and c shown. This model is the 1/8 of the original total model and only contains 9×9×6 gridblocks. In this model, matrix section contains 1 to 6 in XY plane and 2to 6 in Z direction.

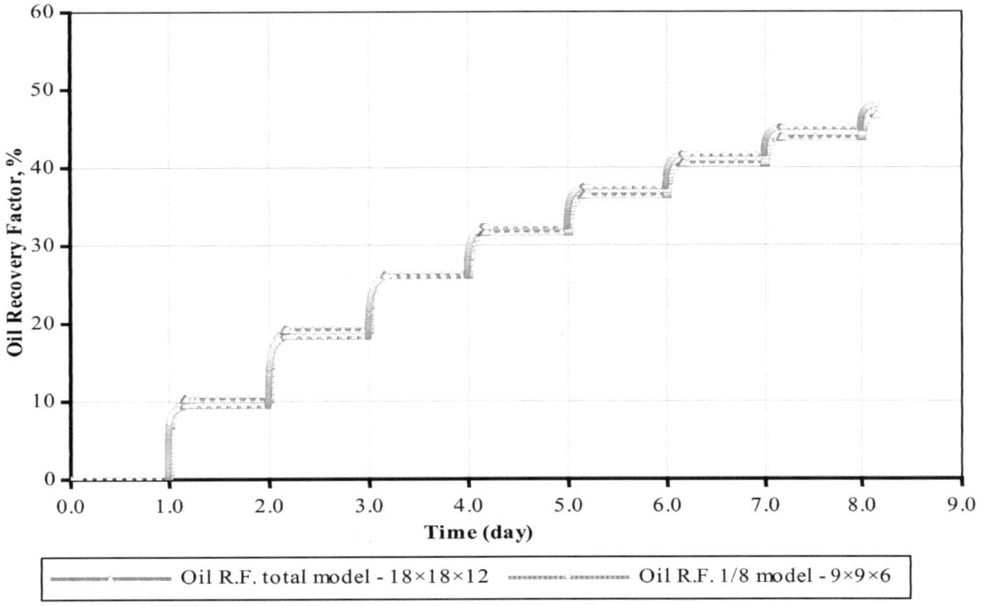

Fig. 6. Difference between the total model and 1/8 model

Fig. 6 compares the difference between the simulation results of the two kinds of models. In the process of seven injection cycles, the difference between them is less than 1.5%, and the 1/8 model only takes 30 minutes to complete the entire process. So we can use this simplified model to conduct our later simulation work with high accuracy instead of the whole model.

Sensitivity Analysis.

Due to the limited data from the cores such as reservoir permeability, porosity and so on, there is high uncertainty in the simulation results. Sensitivity analysis helps us to determine the importance and effect of different parameters on oil recovery.

Usually, there are two different kinds of sensitivity analysis methods. One is called One Parameter at a Time (OPAAT) and the other is Response Surface Methodology (RSM). OPAAT is the method that each parameter is analyzed independently while remaining parameters are set to their reference value. Compared with OPAAT, RSM is a more effective method that multiple parameters are adjusted together then results are analyzed by fitting a response surface to results[15]. In this method, values of multiple parameters are changed simultaneously. The combination of parameter values that are chosen based on an experimental design. The response surface is a proxy for the reservoir simulator that allows fast estimation of the response. The correlation between response and parameters is described in the following equation.

$$NPV = f(x_1, x_2, \ldots, x_n)$$

There are three kinds of polynomial equation to analyze the simulation results.

Liner Model:

$$y = a_0 + a_1x_1 + a_2x_2 + \cdots + a_kx_k$$

Linear and Quadratic Model:

$$y = a_0 + \sum_{j=1}^{k} a_jx_j \;{}_1 + \sum_{j=1}^{k} a_{jj}x_j^2$$

Linear and Quadratic and Interaction Model:

$$y = a_0 + + \sum_{j=1}^{k} a_jx_j + \sum_{j=1}^{k} a_{jj}x_j^2 + \sum_{i<j}\sum_{j=2}^{k} a_{ij}x_ix_j$$

RSM method was used in our analysis through the software of CMOST-CMG. The parameters and the range of values are shown in table 3. Fig.7 shows the resulting tornado plot.

Table 3 Parameters for the sensitivity analysis.

Parameter	Lower Limit	Upper Limit
Bin (methane diffusivity)	0.02	0.15
FracturePerm	50	10000
MatrixPerm	0.00001	0.001
Porosity	0.02	0.1
Pressure	500	10000

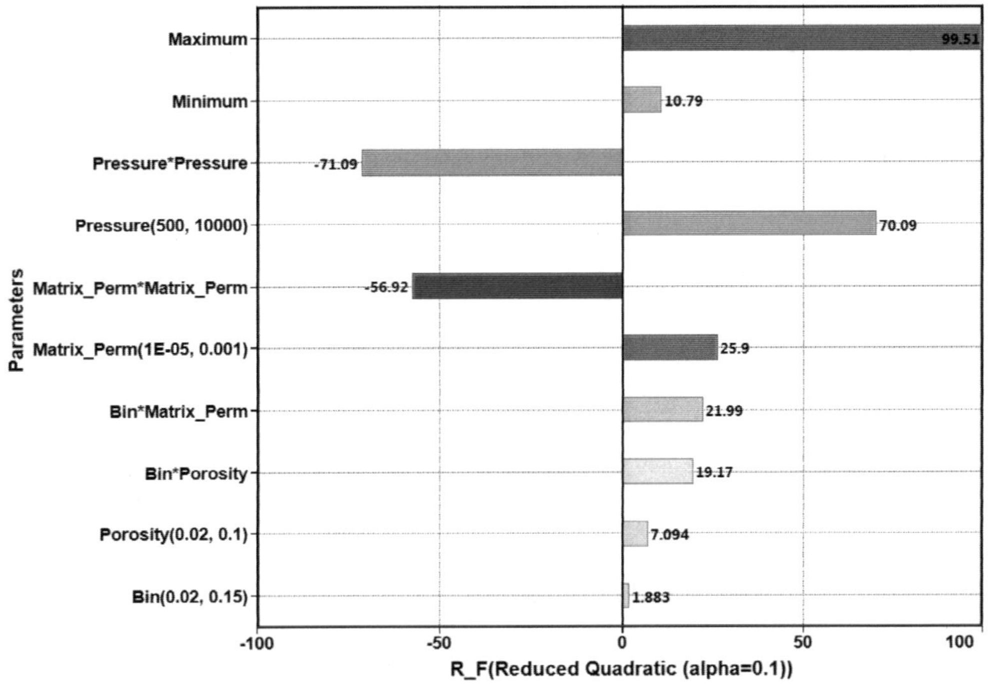

Fig. 7. Resulting tornado plot for sensitivity analysis

This simulation results show that the most important parameter is injection pressure. Increasing pressure from 500 to 10000 psi results in an increase in oil recovery of 70.09 on average. The second important parameter is matrix permeability with an average of 25.9 increased in oil recovery when enhancing the matrix permeability from 0.00005 to 0.001, followed by porosity and methane diffusivity (denoted by bin number). The quadratic term such as P2 means that the parameter has a non-liner relation with the objective functions. The cross term such as Bin×Porosity means that modifying the 2 parameters at the same time has an effect stronger than the sum of their individual liner or quadratic.

History Matching Experiments

When creating a simulation model, there may be uncertainty in the input parameters. History matching is to reduce the error between the simulation results and experiment data to make sure that the model is correct and can be able to predict future behavior. Fig. 8 illustrates the history matching result of oil recovery factor for core with diameter of 1 inch. A good agreement between simulation result and experiment data is achieved. So this model can be used to investigate methane injection in this experiment of different core sizes from Eagle Ford.

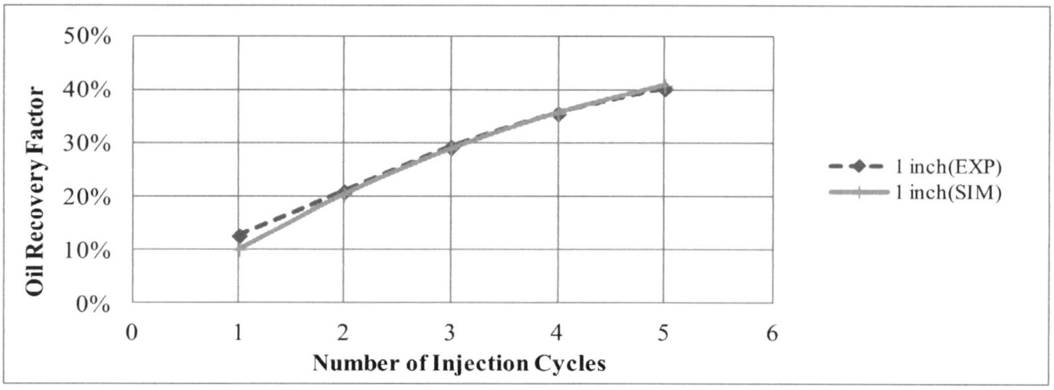

Fig. 8. History matching with core diameter of 1 inch.

Because all the experiments were conducted under the same condition and saturated with the same crude oil, the numerical model can be used to test the results of different core sizes with diameters of 1.5, 2, 2.5, 3, 3.5 and 4 inches respectively. Fig.9 shows the simulation results with different core sizes. The results of simulation are in consistent with experiment data that oil recovery factor decreases with the increase of core diameter.

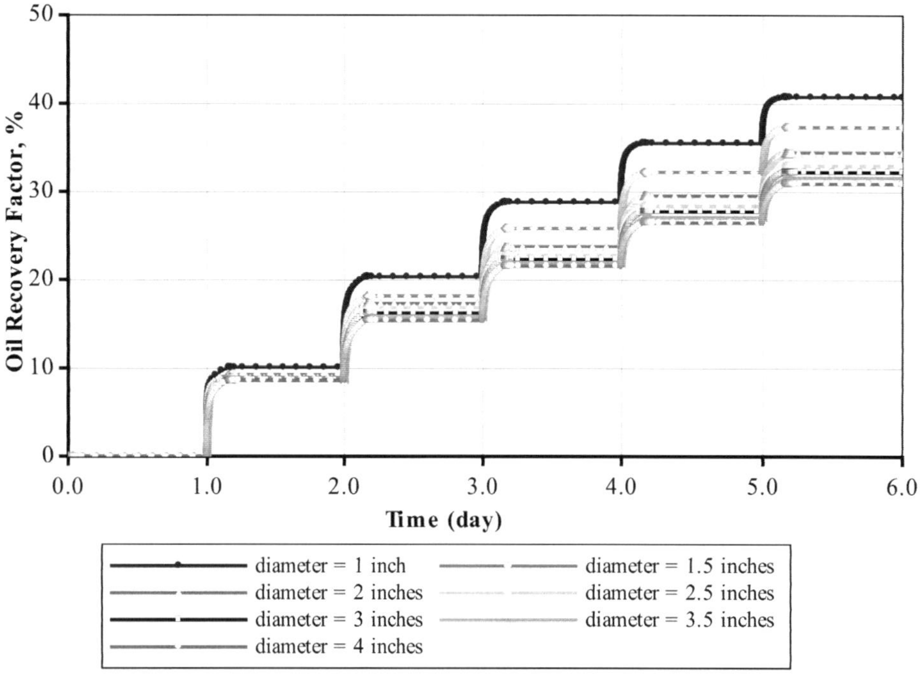

Fig. 9. Simulation results for different core sizes

Upscale Effect

Our current data of the field-scale performance of cyclic methane injection is very limited despite very favorable results obtained in the laboratory-scale experiments. Reservoir simulation is required to properly scale up the process from laboratory to field condition and to understand reservoir data. The simulation model was used to predict the field-scale methane injection situation and the results shows in table 4 and Fig. 10. The laboratory core size is 1inch in diameter and 2 inches in height. In order to simulate the field scale, both the diameter and height of the core were magnified to 150 times and 1500 times in the two enlarged models. As we have illustrated, the oil recovery will decrease with the increase of core size. The results show that the oil recovery reduced by 12.77% when the scale of the core is enlarged from 1 inch to 12.5ft, and decreased by only 0.04% when the size is enlarged increase from 12.5 ft to 125 ft. And in the field-scale the oil recovery after seven cycles is 35.46%. This shows that cyclic methane injection could be effective when applied to the field scale. However, further research is needed on upscaling.

Table 4 Oil recovery data in gas injection cycles compared with upscale model

Core sizes	1	2	3	4	5	6	7
1 inch * 2 inches	10.09	20.33	28.81	35.53	40.79	44.94	48.27
12.5ft * 16.7 ft	8.28	14.72	20.25	25.08	29.05	32.49	35.50
125ft * 167 ft	8.23	14.70	20.21	25.04	29.01	32.45	35.46

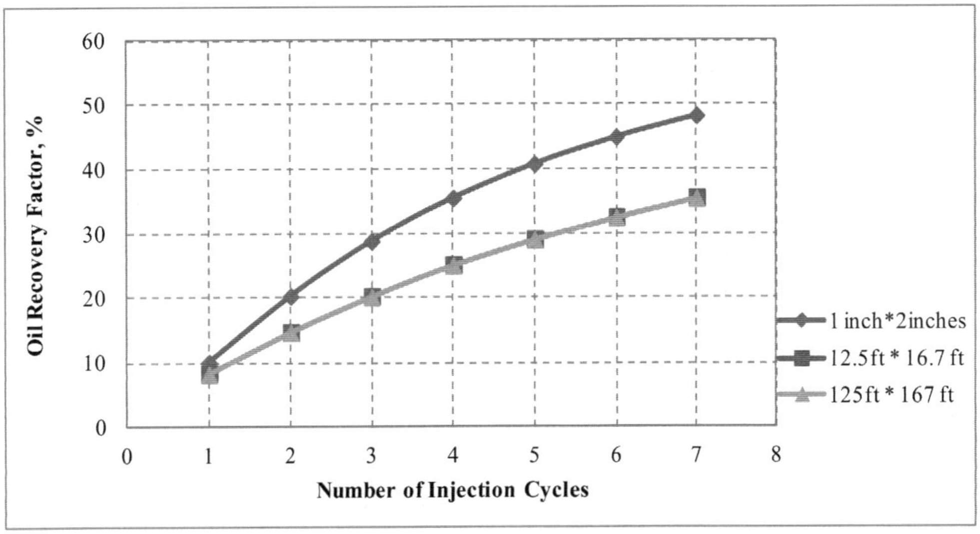

Fig. 10. Oil recovery factor after upscale

Conclusion

In this paper, cyclic methane injection was conducted on shale core plugs from Eagle Ford. The experimental results show that incremental oil recovery in a single cycle decreased with the increase in number of injection cycles. Oil recovery decreased as the core size increased. Sensitivity analysis illustrates that the paramount parameter during cyclic methane injection is injection pressure with an increase of 70.09% in oil recovery on average when the pressure is increased from 500 to 10000 psi. The second important parameter is matrix permeability, followed by porosity and methane diffusivity. Oil recovery decreases about 12.77% when the size in increased from a laboratory-scale to a field-scale.

Acknowledgments

Crude oil and cores were provided by Apache Corporation. The work presented in this paper is supported by the Department of Energy under Award Number DE-FE0024311.

References

1. Gamadi, T.D., Sheng, J.J. and Soliman, M.Y. (2013), "An Experimental Study of Cyclic Gas Injection to Improve Shale Oil Recovery," SPE 166334 in *SPE Annual Technical Conference and Exhibition*, New Orleans, Louisiana, USA, 30 Septembere - 2 October.
2. Gamadi, T.D., Sheng, J.J., Soliman, M.Y., Menouar, H., Watson, M.C. and Emadibaladehi, H. (2014a), "An Experimental Study of Cyclic CO2 Injection to Improve Shale Oil Recovery," SPE 169142 in *SPE Improved Oil Recovery Symposium*, Tulsa, Oklahoma, USA, 12-16 April.
3. Yu, Y. and Sheng, J.J. (2015), "An Experimental Investigation of the Effect of Pressure Depletion Rate on Oil Recovery from Shale Cores by Cyclic N2 Injection," SPE 178494 in *SPE Unconventional Resources Technology Conference*, San Antonio, Texas, 20-22 July.
4. Wang, Z., Ma, J., Gao, R., Zeng, F., Huang,C., Tontiwachwuthikul, P. and Liang, Z. 2013. "Optimizing Cyclic CO2 Injection for Low - permeability Oil Reservoirs through Experimental Study," SPE 167193 in *SPE Unconventional Resources Conference*, Calgary, Alberta, Canada, 5-7 November 2013.
5. Wan, T., Sheng, J.J., Soliman, M.Y. (2013a), "Evaluation of the EOR Potential in Shale Oil Reservoirs by Cyclic Gas Injection," SPWLA d12-00119 in *SPWLA 54th Annual Logging Symposium*, New Orleans, Louisiana, USA, 22-26 June.
6. Yu, W, Lashgari, H., Sepehrnoori, K. (2014a), "Simulation Study of CO2 Huff-n-Puff Process in Bakken Tight Oil Reservoirs," SPE 169575 in *SPE Western North American and Rocky Mountain Joint Meeting*, Denver, Colorado, USA, 17-18 April.
7. Yu, W., Al-Shalabi, E.W., Sepehrnoori, K. (2014b), "A Sensitivity Study of Potential CO2 Injection for Enhanced Gas Recovery in Barnett Shale Reservoirs," SPE 169012 in *SPE Unconventional Resources Conference*, Woodlands, Texas, USA, 1-3 April.
8. Mattax, C. C. and Kyte, J. R. (1962), "Imbibition oil recovery from fractured, water-drive reservoir," *Society of Petroleum Engineers Journal*, 2(02), pp 177-184.
9. Ma, S., Morrow, N. R., Zhang, X. (1995), "Generalized scaling of spontaneous imbibition data for strongly water-wet systems," in *Technical Meeting/Petroleum Conference of The South Saskatchewan Section, Petroleum Society of Canada*, Regina, Canada, 16-18 October.
10. Kewen Li, S., and Roland N. Horne. (2002), "A General Scaling Method for Spontaneous Imbibition," SPE 77544 in *SPE Annual Technical Conference and Exhibition*, San Antonio, Texas, 29 September-2 October.
11. Mojdeh Delshad, N. F. N. (2009), "Scale Up Methodology for Wettability Modification in Fractured Carbonates," SPE 118915 in *SPE Reservoir Simulation Symposium*, Woodlands, Texas, 2-4 February.
12. Pujol, L.; Boberg, T. C. (1972), "Scaling Accuracy of Laboratory Steam Flooding Models," SPE 4191 in *SPE California Regional Meeting*, Bakersfield, California, 8-10 November.

13. Leung, J. Y. and Srinivasan, S. (2011), "Analysis of uncertainty introduced by scaleup of reservoir attributes and flow response in heterogeneous reservoirs," *Society of Petroleum Engineers Journal*, 16(03), pp 713-724.
14. Thomas, S.; Ali, S. M. F.; Thomas, N. H. (1997), "Scale-up Methods For Micellar Flooding And Their Verification," PETSOC 97-50 in *Annual Technical Meeting*, Calgary, Alberta, Canada, 8-11 June.
15. CMG CMOST Manual. (2014), "Proxy Modeling," *Enhance & Accelerate, Sensitivity Analysis, History Matching, Optimization & Uncertainty Version*, PP. 241-261.

Simulation Study on Huff-n-Puff Gas Injection to Enhance Condensate Recovery in Fractured Shale Gas Reservoirs

Xingbang Meng, James J. Sheng, Lei Li, Texas tech University, Lubbock, TX

Abstract

Nowadays, production of liquid dropout -"condensate" in shale gas reservoirs has attained more and more attention. In shale gas reservoirs, when the pressure decreases below the dew point pressure, the heavy part of gas will be dropped out. This liquid dropout is "condensate". The condensate has a very good economic value since it contains the heavier components of initial gas. A critical issue is that the condensate would accumulate near the wellbore and could reduce the productivity of gas. Thus, enhance condensate recovery in shale gas reservoirs could allow producers to make full use of the economic value of condensate, and also it could improve the gas productivity in shale gas reservoirs.

In this paper, a numerical simulation approach is applied to evaluate the enhanced condensate recovery potential in a fractured shale gas reservoir by Huff-n-Puff gas injection. In the simulation work, different types of gas: CO2, and CH4 were injected into the reservoir through different injection schemes. The injection schemes include different injection pressure, different soaking time, and different depletion pressure.

The compositional simulation results indicate that the hydraulic fractures are critical to enhance condensate recovery. Furthermore, by using huff-n-puff gas injection method, the condensate recovery is increased 15%.

Introduction

The production oil and gas resources in shale plays have attained more and more attention. Since in shale gas condensate reservoirs, the condensate is formed by the heavier components of the reservoir fluid. Thus, the condensate has a good economic value. Also, the increased production of condensate would result in the increase of gas production.

Gas condensate reservoirs produce gas in the range of 30 - 300 STB/MMSCF, the ranges of pressure and temperature gas condensate reservoir are between 3000 and 8500 psi and 150 – 400 oF, respectively (Zendehboudi, 2012). Figure 1 presents an example diagram of a gas condensate region.

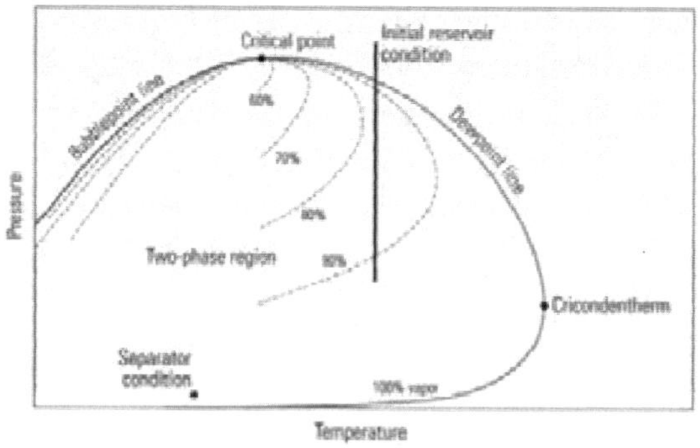

Figure 1: Phase diagram of a retrograde-condensate gas (Li Fan, 2005)

In shale gas condensate reservoirs, after exploration, the pressure of the reservoir would be decreased. Once the pressure is depleted below the dew point pressure, liquid hydrocarbon phase named condensate would be formed. More condensate would be formed as the pressure is continually decreasing. The condensate would accumulate and form condensate blockage. It could reduce the relative permeability of gas and condensate, and the production of gas and condensate would be decreased. The study of Wheaton and Zhang (2000) shows that condensate banking problem is more significant for low permeability condensate systems.

Roussennac (2001) have characterized retrograde gas reservoirs to exhibit three different regions as shown in Figure 2. Region 1 is near the wellbore region, the condensate saturation in this region is higher than the critical saturation. Thus, the condensate is mobile. Region 2 is the part of the reservoir where condensate begins to form because of the pressure depletion. However, since the condensate saturation has not reached critical saturation, this condensate is immobile. Region 3: because the pressure is higher than the dew point pressure, thus, there is only gas exists in this region.

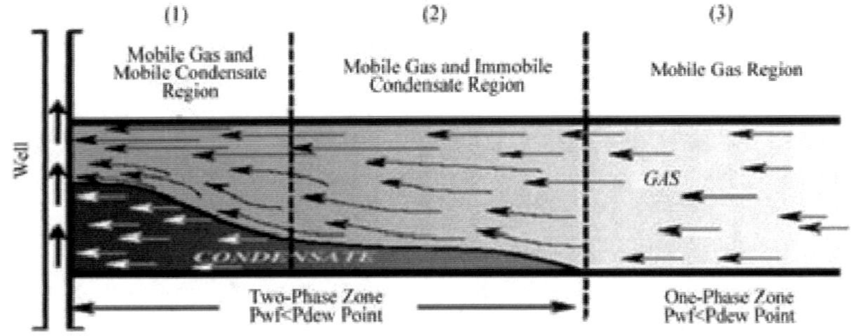

Figure 2: Schematic gas condensate flow behavior in three regions (Roussennac, 2001)

There main way to increase the gas and condensate recovery in gas condensate reservoirs. One is the pressure build up. When the gas is injected into the reservoir, the pressure would be increased to higher than dew point pressure. Sanger and Hagoort (1998) investigated the efficiency of nitrogen to evaporate gas condensate compared to methane. Their study showed that methane can evaporate more condensate than nitrogen. Also, Luo (2000) conducted experiments on a real rich gas condensate fluid to investigate condensate recovery based on the two schemes mentioned above. Their results showed that the condensate recovery is higher when injection is done above the saturation pressure. Also, Amini and, Aminshahidy (2011) did simulation work to enhance condensate recovery by injecting different types of gas: methane, nitrogen and CO2. Simulation results showed that the performance of CO2 is the best.

Huff-n-Puff injection is commonly used in steam injection in conventional reservoir. Both huff-n-puff and gas flooding are ways to increase the reservoir pressure. In huff-n-puff, there is only one well which is used as injection well and production. In huff process, the gas could be injected into the reservoir. Then, the well would be shut in for some time. This is the soaking time. After soaking, the well would be opened again and produce reservoir fluid as a production well. This process is puff. Thus, one cycle including huff process, soaking and huff process.

Actually, huff-n-puff is the best approach among gas injection methods for shale gas condensate reservoirs. As it mentioned, the condensate would accumulate near the wellbore region. Thus, in huff-n-puff method, the condensate would accumulate around the production well. However, this production would be used as the injection well. Thus, the condensate region is just around the injection well. In this way, the pressure of condensate region would be increased much faster. The condensate would be revaporized and when the well is used as production again, the condensate would flow into the wellbore and be recovered. Thus, huff-n-puff is a very good way to enhance the condensate and gas recovery.

Model Setup

A basic Cartesian model and Computer Modeling Group simulator were applied to simulate the huff-n-puff gas injection in shale gas condensate reservoirs. Table 1 shows the gas condensate composition from SPE problem 3. And the related reservoir and fluid parameters are shown in Table 2.

Table 1. Peng-Robinson EOS Fluid Description of Gas Condensate

Components	Initial comp%	Tc(K)	Pc(atm)	ACF	MW g/mole
CO2	0.1618	304.2	72.8	0.225	44.010
C1	0.7098	190.6	45.4	0.008	16.043
C2-C3	0.0790	330.5	45.7	0.119	36.000
C4-C6	0.0260	453.8	34.9	0.226	72.000
C9	0.0200	606.0	26	0.359	128.00
C22	0.0034	869.7	14.9	0.788	310.00

Table 2. Reservoir and fluid properties

Reservoir pressure	5000 psi
Reservoir temperature	335 °F
Matrix permeability	0.0001 md
porosity	0.06

The initial reservoir pressure for this shale gas condensate is 4500 psi. The permeability of the matrix is ultra-tight: 100 nano-Darcy. The implicit assumption for the model setting is that the shale play is homogeneous and isotropic. The shale gas reservoir model is 150-ft (long) * 600-ft (wide) * 200-ft thick. The grid blocks are set as: 10*21*2. One vertical well is set to be used as production well and injection well in huff-n-puff process. Since it is the shale plays, because of the ultra-low permeability, fracture is build neat the well. The fracture permeability is 41.46 md and the fracture width is 2 ft. The model is shown in Figure 3

Figure 3: simulation model

Simulation Results

Effect of huff-n-puff on condensate recovery

In the initial condition of shale gas condensate reservoirs, since the pressure is higher than dew point pressure, there is only gas phase, thus, the oil saturation is 0. After pressure depletion, the condensate would form and the condensate saturation would increase. The condensate would also accumulate near the production well. Then, in huff-n-puff process, when the pressure is increased, the condensate would revaporized and part of them would be recovered in production process. The condensate saturation would be decreased. Figure 4 illustrates the primary depletion and effect of huff-n-puff on condensate saturation very clearly.

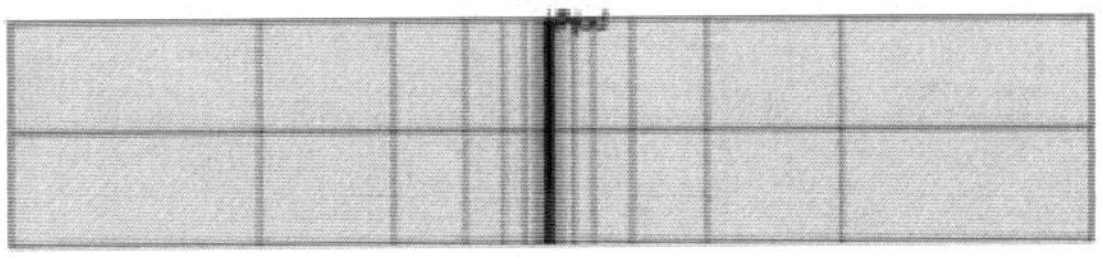

Initial: condensate saturation=0

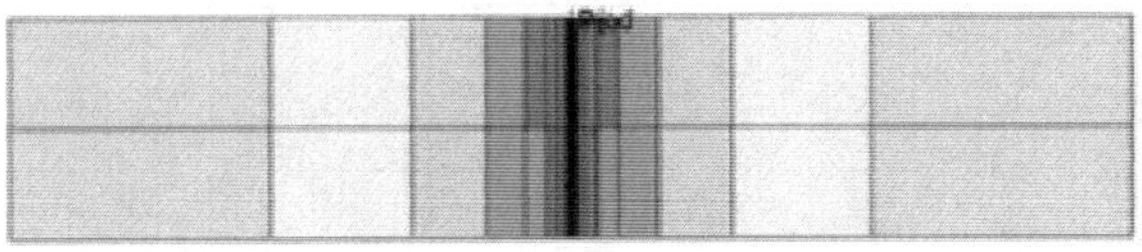

primary depletion: condensate accumulate around near wellbore

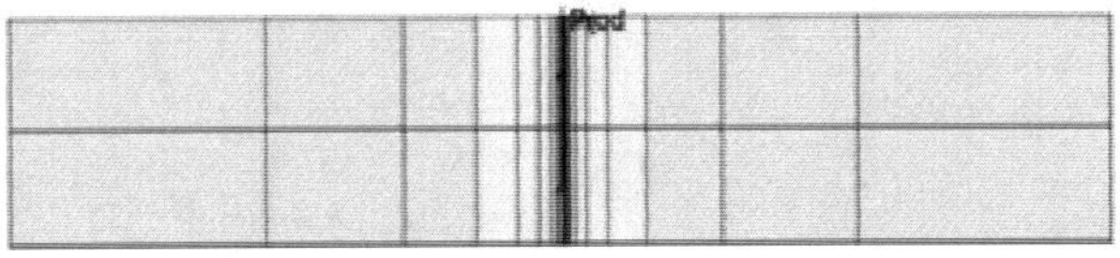

After 1 huff-n-puff: condensate saturation decreases

Figure 4: condensate saturation variation during the process

The primary depletion time is 2800 days. The fracture is set since the beginning. The permeability of shale is so low, if there is no fracture, the fluids would seldom flow out from the reservoir. Figure 5 shows the oil recovery without fracture. We can see that the recovery is extremely low, only 0.045%. Thus, the fracture is added to the model.

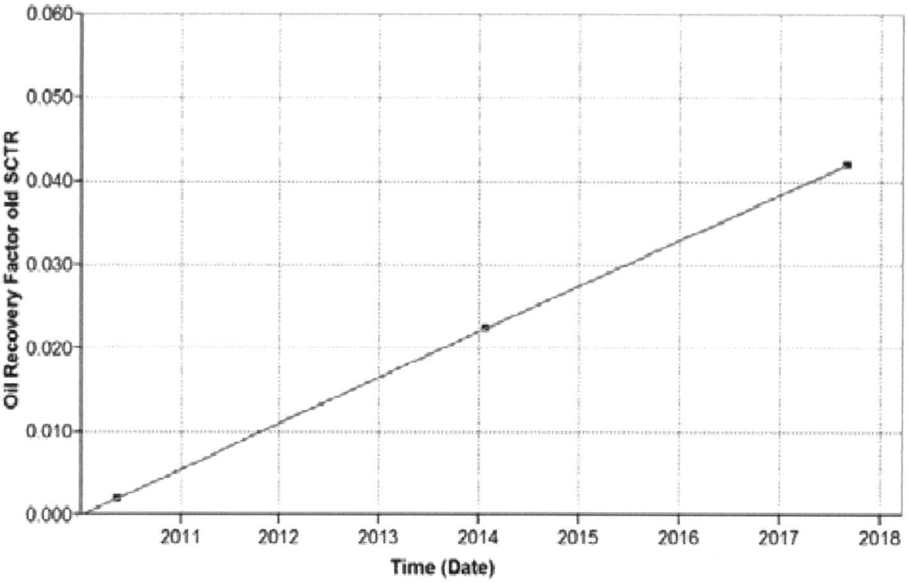

Figure 5: primary depletion without fracture

After primary depletion, the huff-n-puff process is applied. For huff-n-puff, we set injection time 100 days and production time 100 days. The injected gas is methane in the model. Figure 6 shows the pressure and oil recovery for primary depletion and huff-n-puff.

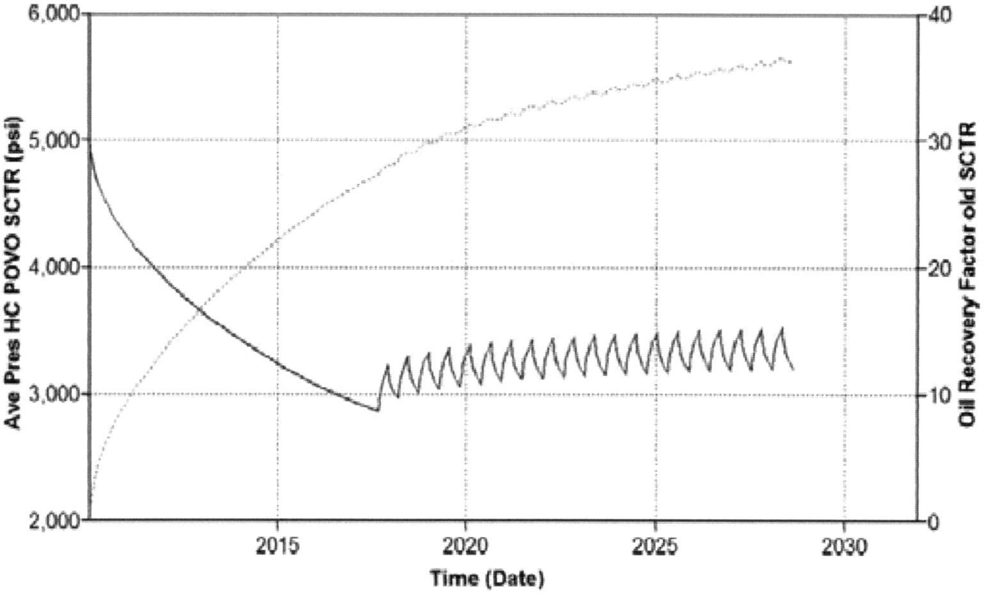

Figure 6: huff-n-puff methane injection, 100 days injection time and 100 days production time

As it shows in figure 6, the primary recovery for fractured shale gas condensate is 25%, after 10 years huff-n-puff injection, the condensate recovery is increased to 37%. It proves that huff-n-puff is an effective way to enhance the condensate recovery in shale gas condensate reservoirs.

Effect of injected gas on condensate recovery

In this simulation model, two different gases is injected into the reservoir: methane and CO2. From the simulation work, we can see that the effect of inject gas is very small in figure 7.

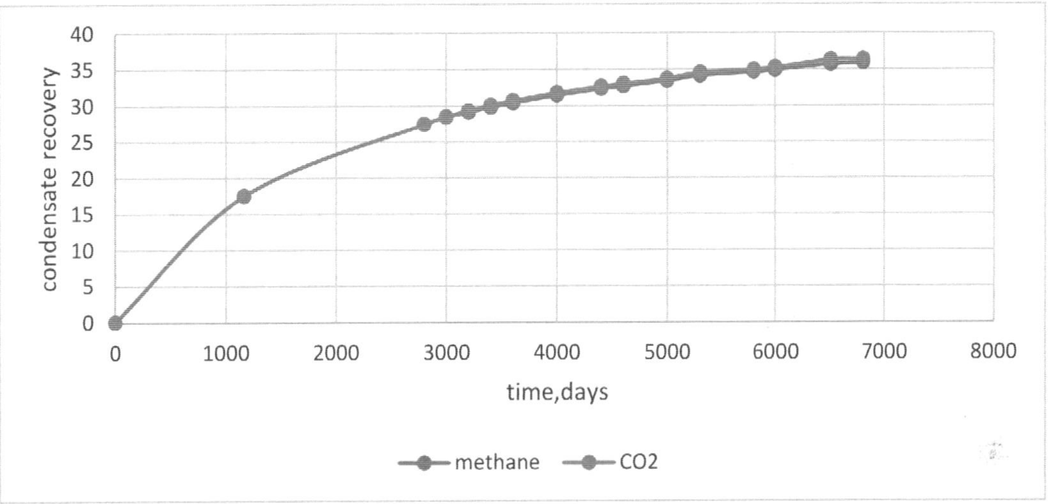

Figure 7: comparison between methane and CO2 injection

Effect of depletion pressure

The simulation model also tests the depletion pressure effect in huff-n-puff process. Three different pressure are tested: 500 psi, 1000 psi, 1500 psi. In primary depletion period, the pressure are same. In huff-n-puff period, the depletion are different. Figure 8 shows the comparison between different depletion pressures. As we can see, the lower of the depletion pressure, higher condensate recovery.

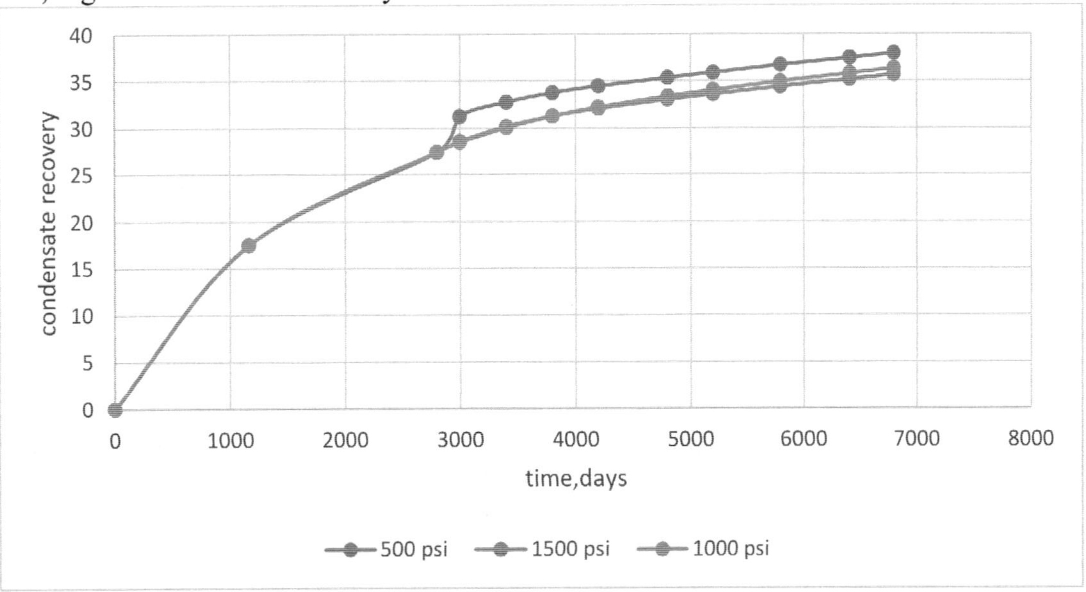

Figure 8: effect of depletion pressure on condensate recovery.

Effect of injection pressure

The different injection pressure is tested. For shale gas condensate reservoirs, the purpose of gas injection is to increase the pressure. Thus, normally, the injection pressure should near or higher than dew point pressure. In this study, two different pressure are tested: 3400 psi and 4500 psi which one is near the dew point pressure and one is higher than dew point pressure. As we can see from the figure 9, the condensate recovery is very similar. This is because when the pressure is higher dew point pressure or near it, the condensate could be revaporized in both cases. Once the condensate is revaporized, it could flow to wellbore with gas mixture.

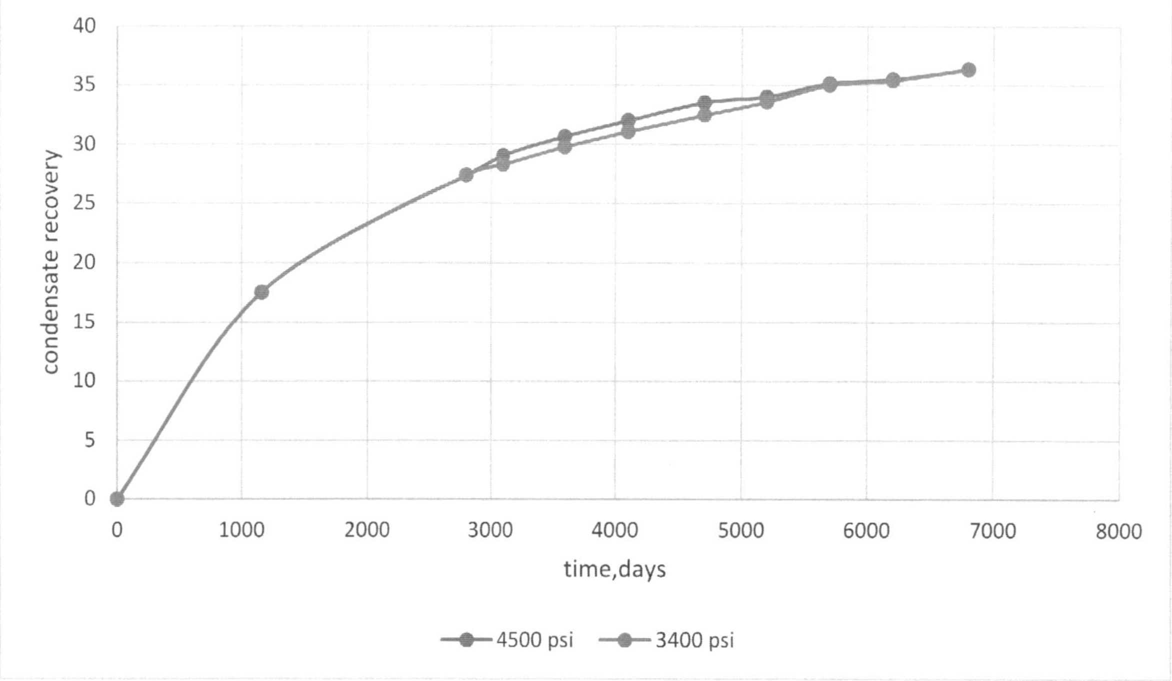

Figure 9: effect of injecting pressure on condensate recovery

Conclusions

1. The simulation study proves that huff-n-puff is an effective way to enhance condensate recovery in fractured shale gas condensate reservoirs. It could increase pressure of condensate region very fast and vaporize the condensate.

2. The efficiency of injecting C02 is almost same as injecting methane. The injecting gas type does not affect the recovery.

3. The lower depletion pressure, the higher condensate recovery.

4. When the injection pressure is near or higher than dew point pressure, the incremental of pressure would not have much condensate recovery incremental. Thus, in real reservoir exploration, the injection pressure could be set just near the dew point pressure.

References

1. Amini, Sh., Aminshahidy, B., Afshar, M. (2011), "Simulation Study of Enhanced Condensate Recovery in a Gas-Condensate Reservoir". Iranian Journal of Chemical Engineering, Vol. 8, No. 1.

2. Luo, K., Li, S., Zheng, X., Chen, G., Dai, Z., Liu, N. (2001), "Experimental Investigation into Revaporization of Retrograde Condensate by Lean Gas Injection", paper SPE 68683-MS, presented at SPE Asia Pacific Oil and Gas Conference and Exhibition, 17-19 April, Jakarta, Indonesia.

3. Roussennac, B. (2001), "Gas Condensate Well Test Analysis (master thesis). Stanford University.

4. Sanger, P. J., Hagoort, J. (1995), "Recovery of Gas-Condensate by Nitrogen Injection Compared with Methane Injection." Society of Petroleum Engineers, SPE-30795-PA.

5. Wheaton, R.J., Zhang, H.R. (2000), "Condensate Banking Dynamics in Gas Condensate Fields: Compositional Changes and Condensate Accumulation around Production Wells", paper SPE 62930-MS, presented at SPE Annual Technical Conference and Exhibition, 1-4 October, Dallas, Texas.

6. Zendehboudi, S., Ahmadi, M.A., James, L., Chatzis, I. (2012), "Prediction of Condensate-to-Gas Ratio for Retrograde Gas Condensate Reservoirs Using Artifical Neural Network with Particle Swarm Optimization." Energy & Fuels, 26(6), 3432-3447.

442933 Downstream Challenges Processing Opportunity Crude Oils

Abstract:

Tight oil availability has transformed the refining industry in the U.S. Rapidly increasing availability of tight oil is an opportunity but has handling and processing issues. A brief summary of the characteristics and variability of tight oil will be given. Examples of operational challenges will be presented.

An Experimental Study of the Potential of Improving Shale Oil Recovery By Gas Flooding

Yang Yu and James J. Sheng
Texas Tech University

15AIChE
Annual Meeting, Salt Lake City, UT

TEXAS TECH UNIVERSITY

Outline

- Introduction
- Experimental Work
- Results and Discussion
- Conclusion
- Acknowledgements

15AIChE
Annual Meeting, Salt Lake City, UT

TEXAS TECH UNIVERSITY

Introduction

- Tight oil production leads the total oil production in US in recent years
- As the primary recovery factor remains low, we will face the challenge of improving shale oil recovery
- Gas injection is a promising EOR technique in shale reservoirs
- The recovery performance of huff-n-puff process have been investigated experimentally, while limited gas flooding tests conducted using shale samples

15AIChE
Annual Meeting, Salt Lake City, UT

TEXAS TECH UNIVERSITY

Introduction

Objective:

- Evaluate the shale oil recovery potential by using gas flooding
- Using CT scanner to monitor the experiment process and analyze the production history

15AIChE
Annual Meeting, Salt Lake City, UT

TEXAS TECH UNIVERSITY

Experimental Work

Materials

- Core type: Wolfcamp outcrop, 1.5" Diameter and 2" Length
- Gas source: Nitrogen gas
- Saturated fluids: Crude oil from Wolfcamp Shale

Table 1. Crude Oil Properties (21 °C)

Density (g/cm^3)	Viscosity (cp)	Surface Tension (dynes/cm)	CT Number (Hounsfield)
0.85	8.3	25	-188

15AIChE
Annual Meeting, Salt Lake City, UT

TEXAS TECH UNIVERSITY

Experimental Work

Table 2. Operational parameters

Oil Saturation pressure	1,000 psig
Injection pressure	1,000 psig
Outlet pressure	0 psig
Confining pressure	1,500 psig
Flooding time	1-day, 3-day, and 5-day
Operation temperature	21°C

15AIChE
Annual Meeting, Salt Lake City, UT

TEXAS TECH UNIVERSITY

Experimental Work

Procedures:

1. Dry and weigh core samples
2. Vacuum and saturate samples with crude oil
3. Scan the plug before and after the saturation process
4. Place the core in core holder and apply confining pressure
5. Apply inlet pressure and begin N2 flooding
6. Monitor the recovery process throughout the experiment using CT

15AIChE
Annual Meeting, Salt Lake City, UT

TEXAS TECH UNIVERSITY

Experimental Work

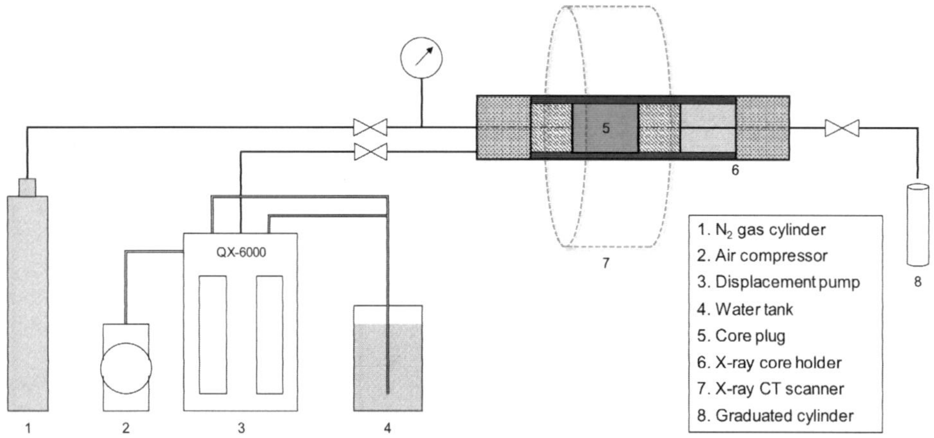

Figure 1. Experimental setup for gas flooding

15AIChE
Annual Meeting, Salt Lake City, UT

TEXAS TECH UNIVERSITY

Results

Recovery factor (RF)

$$Weight\ loss:\quad RF = \frac{W_{sat} - W_{exp}}{W_{sat} - W_{dry}}$$

$$CT\ Number:\quad RF = S_{N2} = \frac{CT_{sat} - CT_{exp}}{CT_{sat} - CT_{dry}}$$

15AIChE
Annual Meeting, Salt Lake City, UT

TEXAS TECH UNIVERSITY

Results

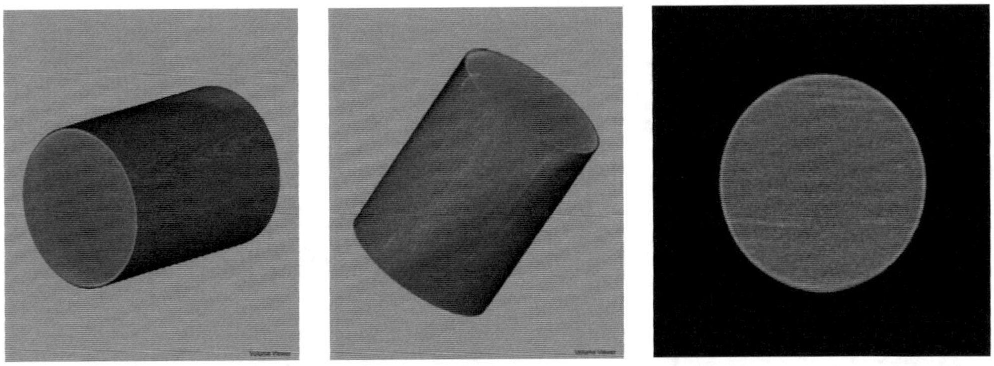

Figure 2. CT image in 3D and slice view

*No natural factures/fissures show

15AIChE
Annual Meeting, Salt Lake City, UT

TEXAS TECH UNIVERSITY

Results

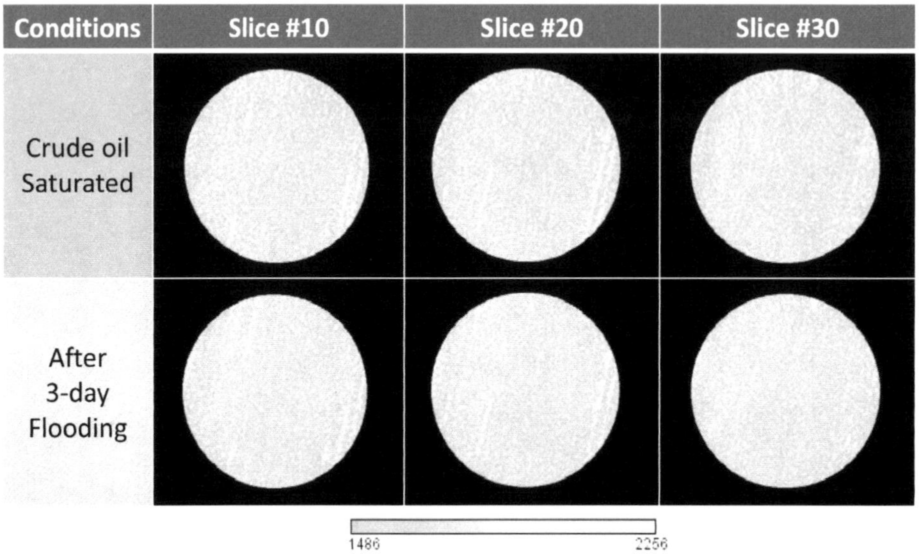

Figure 3. Colored CT images of GF-W#2 before and after N2 flooding process

More area in blue color: the average CTN was decreased after flooding

15AIChE
Annual Meeting, Salt Lake City, UT

TEXAS TECH UNIVERSITY

Results

Core No.	Flooding time (Days)	Dry weight (g)	Sat. weight (g)	Porosity	RF
GF-W#1	1	123.7400	128.7890	10.76%	17.18%
GF-W#2	3	123.6614	127.8540	8.94%	29.74%
GF-W#3	5	124.6270	129.4628	10.31%	39.42%

15AIChE
Annual Meeting, Salt Lake City, UT

TEXAS TECH UNIVERSITY

Results

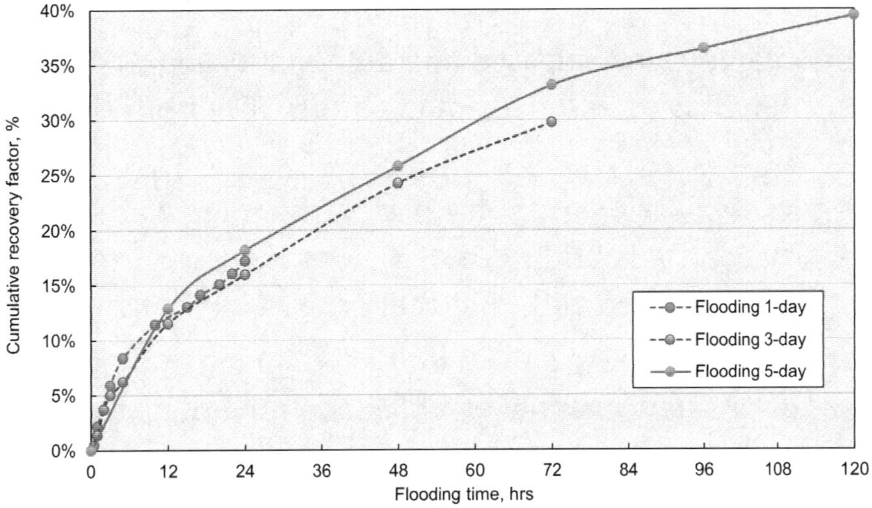

Figure 4. Recovery factor vs. flooding time

15AIChE
Annual Meeting, Salt Lake City, UT

TEXAS TECH UNIVERSITY

Discussions

- After one day's recovery, 16-18% of oil was produced; four more days flooding period contributed 20% more production
- Production rate was high until gas breakthrough achieved
- Once the fluid flow path was established, the injected gas would flow through the limited communication channels thus oil production rate dropped down gradually.
- Effect of injection pressure on RF will be evaluated in further study.

15AIChE
Annual Meeting, Salt Lake City, UT

TEXAS TECH UNIVERSITY

Conclusions

- Experimentally investigated and demonstrated the potential of gas flooding to improve liquid oil recovery from shale reservoirs
- The best recovery performance of N2 flooding was shown in the initial period, about 50% of oil was yielded during one-day recovery compared with the total production of five days.
- High oil recovery rate shown in the beginning and maintained until the gas flow channel established in the plug, then such rate was decreased with the increase of flooding time.

15AIChE
Annual Meeting, Salt Lake City, UT

TEXAS TECH UNIVERSITY

Acknowledgements

U.S. DEPARTMENT OF ENERGY

The work presented is supported by the Department of Energy under Award Number DE-FE0024311.

15AIChE
Annual Meeting, Salt Lake City, UT

TEXAS TECH UNIVERSITY

Questions/Comments

Thank you!

15AIChE
Annual Meeting, Salt Lake City, UT

TEXAS TECH UNIVERSITY

422843 Conversion of Biomass-Derived Isoprene and Acrolein to Terephthalate Acid (TPA) and Hydrocarbon Fuels

Abstract:

The conversion of biomass to biofuels and value-added chemicals has attracted numerous attentions in terms of less dependence on fossil fuel and increasing concerns on environmental problems. Hydrocarbon fuels such as diesel and jet fuel (C_8-C_{21}) have an advantage because of compatibility with the current infrastructure and a higher energy density suitable for truck and jet engines. Terephthalic acid (TPA) is an indispensable monomer precursor for polyesters such as polyethylene terephthalate (PET), a thermoplastic polymer with a broad spectrum of applications, ranging from drink bottles to polyester fibers. Several routes have studied to produce either TPA or hydrocarbon fuels using sustainable chemical blocks (e.g. furfural) as starting materials. Herein, we reported new approaches for the synthesis of both TPA and hydrocarbon fuels (C_8 and C_{16}) using abundant biomass-derived chemicals isoprene and acrolein as the starting materials. TPA can be prepared by a simple route avoiding the *p*-xylene production including Diels-Alder, aromatization and oxidation with a total yield of 32%. Then, by sharing the same intermediate, C_8 hydrocarbon can be prepared by a hydrodeoxygenation reaction from the intermediate 4-methyl-cyclohex-3-enecarbaldehyde of Diels-Alder reaction with up to 57% total yield. The transition metal catalysts such as Pt/C, Pd/C, Fe/SiO_2 and Ni/SiO_2 were screened to optimize the product yield. Furthermore, a long chain C_{16} hydrocarbon was also synthesized by a sequence of Diels-Alder reaction, reductive coupling, and hydrodeoxygenation from isoprene and acrolein as well.

435491 Role of Calcination Temperature on the Hydrotalcite Derived MgO-Al2O3 in Converting Ethanol to Butanol

Abstract:

The projected crude oil depletion along with increased greenhouse gas emissions has created great interest in developing technologies to produce fuels and chemicals from renewable resources. Ethanol can be generated efficiently from renewable resources and it is expected to be in surplus due to the limitation on the utilization of ethanol in the gasoline blends. Ethanol condensation chemistry provides an excellent opportunity to convert ethanol to high value building block compound butanol which can be easily converted to fuel or chemical compound. The conversion of ethanol to butanol occurs via a complex network of reaction over a bifunctional acid base catalytic system. The hydrotalcite [$Mg_6Al_2(CO_3)(OH)_{16}$•$4H_2O$] derived MgO-Al_2O_3 mixed oxide catalyst is identified as one of the efficient material to convert ethanol to butanol. The catalyst preparation method, catalyst composition and the experimental conditions plays a role in directing the ethanol condensation chemistry. Our experimental results show a good correlation between the catalyst calcination temperature and the catalyst activity to the ethanol condensation reaction. The catalyst characterization results show that the change in acid base properties with respect to the catalyst calcination temperature is responsible for the change in the catalyst activity. In this talk the experimental and catalyst characterization results will be discussed in detail to show the role of calcination temperature on the hydrotalcite derived MgO-Al_2O_3 in converting ethanol to butanol.

431430 Selective Hydrogenation of Furfural to Furfuryl Alcohol over Supported Pd-Re Catalysts

Abstract:

$PdRe/Al_2O_3$ catalysts are demonstrated to be highly active and selective for hydrogenation of furfural to furfuryl alcohol (FAL). Bimetallic PdRe catalysts (1:1 atomic ratio) were prepared by incipient wetness co-impregnation of γ Al_2O_3 and by two sequential impregnation methods: (1) impregnation of Re/Al_2O_3 with Pd^{2+} with subsequent calcination and (2) impregnation of reduced and passivated Pd/Al_2O_3 with perrhenic acid, [$HReO_4$], without subsequent calcination. Catalyst structure after *in situ* reduction at 400°C depends strongly on the Pd precursor and the sequence of metal precursor addition. Extended x-ray absorption fine structure (EXAFS) spectroscopy and temperature-programmed hydride decomposition (TPHD), showing essentially complete suppression of β-PdH_x formation, evidence alloy formation in catalysts prepared by co-impregnation. Due to strong electrostatic adsorption of $[Pd(NH_3)_4]^{2+}$ at pH 10-11, catalysts prepared from $[Pd(NH_3)_4(NO_3)_2]$ contain small (<5 nm) Pd particles, whereas $[Pd(NO_3)_2]$-derived catalysts contain larger (>10 nm) particles. The latter observation correlates with formation of PdO crystallites during calcination, as detected by Raman spectroscopy. Preparation of $PdRe/Al_2O_3$ catalysts by sequential impregnation facilitates low-temperature Re reduction and increases furfural hydrogenation activity and FAL selectivity. Varying degrees of Pd-Re interaction in each catalyst are indicated by EXAFS spectroscopy, scanning transmission electron microscopy with energy dispersive x-ray analysis (STEM-EDX), and CO diffuse reflectance infrared Fourier transform spectroscopy (DRIFTS). Furfural turnover frequency and FAL selectivity correlate strongly with suppression of H_2 chemisorption; a H/CO chemisorption ratio of ~0.2 appears to be optimum for selective hydrogenation of furfural to FAL.

421112 Solvation Dynamics and Energetics of Hydride Transfer Reactions in Cellulosic Biomass Conversion

Abstract:

Cellulose and its sugar derivatives need to be dissolved in a solvent for their catalytic conversion to chemicals and fuels intermediates. Experimental studies have shown that the addition of a co–solvent to water or replacing water with an alternate solvent can significantly alter conversion and selectivities in these reactions. [1] It is believed that solvents can play dual role in biomass reactions. (i) A solvent can either preferentially solvate an "active" functional group of a biomass molecule or its derivative and protect it from taking part into the reaction [2], or (ii) on the other hand, solvent can also directly participate in the reaction.[3] Hydride transfer is a commonly observed, and usually the rate limiting step in some of the key biomass reactions like isomerization and dehydration.[4] Solvent dynamics and non–equilibrium solvation can play a major role in altering the energetics of charge transfer steps like hydride transfer.[5,6] In the present paper, we perform Car–Parrinello molecular dynamics-metadynamics simulations of intramolecular hydride transfer in glucose molecule, as a model system, in the presence of explicit quantum mechanically treated solvent molecules. Free energy change and activation free energy barriers for the hydride transfer step are computed as a function of solvent composition. It is observed that hydride transfer can be either exergonic or endergonic in different solvents. Similarly the free energy barrier for hydride transfer is a strong function of the solvent environment. Analyses of changes in the electronic structure of the reacting system and in solvent orientation along the trajectory suggest that changes in the charge structure of the sugar molecule polarizes the solvent and that hydride shift takes place in a non–equilibrium solvation environment. Orientational relaxation of the solvent, after the hydride transfer, is observed to be a significantly slower process. The free energy barrier is a result of non–equilibrium solvation and the exergonicity or endergonicity of the hydride transfer step depends on the relaxation dynamics of the solvent.

References:

1. Y. Román-Leshkov et al., *Science*, **2006**, 312, 1933-1937.
2. S. H. Mushrif et al., *Phys Chem Chem Phys*, **2012**, 14, 2637-2644.
3. G. Li et al., *Catalysis Science & Technology*, **2014**, 4, 2241-2250.
4. R. Bermejo-Deval et al., *P Natl Acad Sci USA*, 2012, 109, 9727-9732
5. S. H. Mushrif et al., *Chem. Eng. Sci.* **2015**, *121, 217-235.*
6. *S. H. Mushrif et al., Phys Chem Chem Phys,* ***2015****, 17, 4961-4969.*

Hydrothermal Treatment of Protein, Polysaccharide, and Lipids Alone and In Mixtures

Terigule, Ligang Luo[1], Phillip E. Savage[2,*]

Department of Chemical Engineering, University of Michigan, Ann Arbor, MI, 48109

[1]Present Address: Department of Chemistry, East China Normal University, Shanghai 200241, PR China

[2]Present Address: Department of Chemical Engineering, Pennsylvania State University, University Park, PA 16802

Abstract

We subjected a set of model compounds (cornstarch and cellulose as model polysaccharides, soy protein and albumin as model proteins, sunflower oil and castor oil as model lipids) to the processing conditions and product recovery protocol commonly used for hydrothermal liquefaction of algal biomass to make crude bio-oil. The model compounds were treated individually and in mixtures at 300 and 350 °C for batch holding time ranging from 10 to 90 min. The model lipids produced the highest yield (> 90 wt %) of biocrude (operationally defined as material soluble in dichloromethane), followed by the model proteins (~ 30 – 35 wt %) and then the model polysaccharides (~ 10 – 15 wt %). The production of biocrude at 350 °C occurred fully within the first ten minutes of treatment, and the biocrude yield changed very little at longer times. Liquefaction at 350 °C and 60 min nearly doubled the biocrude yields from polysaccharides relative to those obtained at 300 °C and 20 min. Otherwise, the yields from the different model compounds at the milder and the more severe conditions were comparable. In most instances, the biocrude yield from hydrothermal treatment of mixtures was very similar to the mass-averaged yield calculated from the individual compound results. The chief exceptions were binary combinations of polysaccharide and protein at the more severe conditions. For these mixtures, the biocrude yield exceeded the mass-average yield calculated from the pure compound results, thereby providing evidence that interactions influencing the biocrude yield can occur during hydrothermal treatment of mixtures of the biomolecules. Even so, a quantitative model built on the assumption that the lipids, polysaccharides, and proteins react independently during HTL predicted biocrude yields for ternary mixtures more accurately than did a model with three additional parameters that allowed for the possibility of interactions between the different model compounds.

425067 Catalytic Oxidation of Lignin for the Production of Value-Added Aromatic Aldehydes

Abstract:

Catalytic wet air oxidation (CWAO) of lignin or phenolic composition in waste water is a useful pathway for the recovery of valuable compounds. In this study, bamboo lignin was catalytically oxidized to aromatic aldehydes and acetoderivatives, so was acetovanillone oxidized to vanillin. A total products yield of 9.5% from bamboo lignin, and a vanillin yield of 51% with greater than 90% selectivity from acetovanillone were achieved. The proposed reaction pathways suggest a two-step route towards the formation of value-added aromatic aldehydes from lignin via degradation of acetoderivatives. The kinetics study for the degradation of acetovanillone and formation of vanillin were reported for the first time over the temperature range of 120-150 °C, with the activation energies of 85.29 kJ/mol and 120.7 kJ/mol respectively. To date, CWAO breaks down the lignin polymer and toxic phenolic compounds reasonably and effectively, producing value-added aldehydes, which has a great potential to becoming a new route for the recovery of value-added products from lignin.

434800 Effect of Redox Properties of LaCoO3 Perovskites on Production of Lactic Acid from Cellulosic Biomass

Abstract:

Effect of Redox Properties of $LaCoO_3$ Perovskites on Production of Lactic Acid from Cellulosic Biomass

Xiaokun Yang, Lisha Yang, Hongfei Lin*

Department of Chemical and Materials Engineering, University of Nevada,

1664 N. Virginia St., Reno, NV 89557, USA

E-mail: HongfeiL@unr.edu

Cost-effective conversion of cellulosic biomass to value-added lactic acid with heterogeneous catalysts has attracted much attention recent years. While both solid Lewis acids and bases are extensively studied, which exhibit good catalytic performance but are unfortunately unstable in hydrothermal media, the role of redox catalysts for the production of lactic acid is barely understood. Herein, the $LaCoO_3$ perovskite metal oxides with strong redox properties and a good stability in hydrothermal media were used as the catalyst for the conversion of a variety of cellulosic biomass to lactic acid. Different gas atmosphere including oxidative, inert and reductive gases had the profound effect on the yield of lactic acid. The effects of reaction conditions such as reaction temperature and catalyst loading were also investigated. At the optimum conditions, the yields of ~40%, 38%, and 24% lactic acid were obtained from glucose, xylose and cellulose, respectively. The key intermediates and final products were used as the probe reactants to explore the reaction mechanism. Unlike Lewis acid or base catalysed biomass conversion reaction pathways, our proposed redox reaction pathway has the distinct characteristic that lattice oxygen atoms in the $LaCoO_3$ perovskite structure participate the redox cycles in the sugar conversion to lactic acid. The stability of the $LaCoO_3$ catalyst was also discussed.

435331 H2 Volume Generation from Thermochemical Water-Splitting Reaction Using Core-Shell and Immobilized Ferrite Nanomaterials

Abstract:

During multiple thermochemical water-splitting cycles for hydrogen generation, redox materials undergo sintering or grain growth. Consequently, stable hydrogen volume generation is not realized. To address this issue, we report core-shell nanoparticle morphology of redox materials, which is anticipated to prevent grain growth. Core-shell nanoparticles were synthesized using sol-gel assisted surfactant templating method. In particular, non-ionic and ionic surfactants were explored for the synthesis of ferrite nanoparticles with porous ceramic shell. The core-shell ferrite nanoparticles were analyzed by X-Ray Diffraction (XRD) and Scanning Electron Microscopy (SEM) and Transmission Electron Microscopy (TEM). TEM images confirmed the presence of core-shell morphology with ferrite nanoparticles residing inside the porous ceramic shell. These nanoparticles were loaded inside the Inconel tubular reactor as a packed-bed and regenerated for 2 hours at 1100 deg C under the nitrogen flow. Water-splitting step was also performed at the same temperature of 1100 deg C. To compare sintering or grain growth, ferrite nanoparticles were also immobilized on ceramic support and these materials were investigated for hydrogen generation at water-splitting and regeneration temperatures of 1100 deg C. The hydrogen volume generated by the core-shell nanoparticles was appeared to be relatively similar during multiple thermochemical cycles, however, it was found to be lower as compared to ferrite nanoparticles. The results obtained on the hydrogen volume generated by core-shell nanoparticles and immobilized ferrite nanostructures during multiple thermochemical cycling will be presented.

436714 Heterostructured Cu2O/CuO Decorated with Nickel As a Photocathode for Photoelectrochemical Water Splitting

Abstract:

The design, synthesis and characterization of a novel Cu_2O/CuO heterojunction decorated with nickel cocatalyst as a highly efficient photocathode for solar hydrogen production was investigated. Due to the heterojunction synergism, the Cu_2O/CuO gave a remarkably improved photocurrent density (-2.1 mA/cm^2), i.e. 3.1 times higher than a Cu_2O photoelectrode. Additionally, the Cu_2O/CuO heterojunction, when decorated with nickel cocatalyst, showed six-fold and two-fold increases in photocurrent density (-4.3 mA/cm^2) respectively when compared to Cu_2O and bare Cu_2O/CuO at 0 V vs. RHE. Interestingly, the surface modification with Ni not only effectively promotes water splitting but also stabilizes the photoelectrode. The enhanced photoelectrochemical performance is attributable to the efficient charge transfer and protective role of Ni, the improved crystallinity and the synergism effect of the heterojunction in light absorption and charge separation. This inexpensive photocathode with increased photocurrent density and photostability offers a higher promise for solar hydrogen production.

جامعة الملك فهد للبترول والمعادن
King Fahd University of Petroleum & Minerals

15 AIChE Annual Meeting, Salt Lake City, UT

MULTI-OBJECTIVE OPTIMIZATION OF MEMBRANE MATERIALS SELECTION FOR DIRECT METHANOL FUEL CELL SYSTEM DESIGN

J. K. Adewole[1], A. S. Sultan[1], A. Al-Ahmed[2], S. M. J. Zaidi[3]

[1]Center for Petroleum & Minerals, King Fahd University of Petroleum & Minerals, Dhahran 31261, Saudi Arabia

[2]Center of Research Excellence in Renewable Energy, King Fahd University of Petroleum & Minerals,
Dhahran 31261, Saudi Arabia

[3]School of Chemical Engineering, The University of Queensland, Australia

Outlines

- Introduction
- Objectives
- Experimental Procedure
- Results and Discussion
- Conclusion and Recommendations
- References
- Acknowledgement

KFUPM

Introduction

- Growing need for development of low emission power sources
- Research interest on fuel cell technology for portable, stationary and automotive devices
- The proton exchange membrane (PEM) is one of the primary components of a fuel cell devices
- Good PEM material has the following requirements
  - high proton conductivity
  - low methanol crossover
  - high ion exchange capacity
  - tolerant to water fluctuation in high temperature operations

KFUPM

Introduction

- Previously, PEM materials are selected based on intuitional guidelines
- Today, there are quite a large number PEM materials
- Many performance properties for material selection
- Model based materials selection methodology becomes necessary
- This is thus the main focus of the present study

KFUPM

© King Fahd University of Petroleum & Minerals | Dhahran, 31261 | Saudi Arabia | +966 (3) 860 0000

Main Objective

- To demonstrate a model based method for optimizing the PEM material selection

KFUPM

Experimental Procedure

- Membrane preparation by solvent casting method
- Measurement of
  - methanol permeability
  - water uptake
  - ion exchange capacity

KFUPM

© King Fahd University of Petroleum & Minerals | Dhahran, 31261 | Saudi Arabia | +966 (3) 860 0000

Modeling and Optimization Procedure

- Cost per unit mass was calculated using the linear mixing rule

$$C_{mix} = C_1 w_1 + C_2 w_2 \quad \text{..............(1)}$$

- Optimization was done using multi - objective optimization
- Effect of the relative importance of each PEM property was incorporated using weighted properties approach (WPA)

KFUPM

Modeling and Optimization Procedure

- The weighting factor used for the WPA was determined using the modified digital logic method (MDL)
- The objectives are described as follows:
  - Low methanol (crossover) permeability: P_{MeOH}
  - Low water uptake: W_u
  - High ion exchange capacity: I_C
  - Low Cost of material: C_M

KFUPM

Modeling and Optimization Procedure

- The multi-objective model was defined as:

$$\max(I_C) \qquad \text{.....................(2a)}$$

$$\min(P_{MeOH}, W_u, C_M) \qquad \text{..............(2b)}$$

KFUPM

Modeling and Optimization Procedure

- Property scaling
  - For properties that need to be maximized:

$$Y_i = \frac{X_i}{X_{max}} \times 100 \qquad \text{..............(3)}$$

  - For property that need to be minimized:

$$Y_i = \frac{X_{min}}{X_i} \times 100 \qquad \text{..............(4)}$$

Y_i is the scaled membrane properties, X_i is the numerical value of the property,

X_{max} is the maximum value in the list of the property

X_{min} is the minimum value in the list of the property

KFUPM

Modeling and Optimization Procedure

- **The objective function (performance index)**

$$\gamma = \sum_{i=1}^{n} Y_i \alpha_i \qquad \text{..............(5)}$$

i is summed over all the n relevant properties,

γ is the material performance index and

α is the weighting (scaling) factor

Results and Discussion

Table I. Performance Properties of Various Composites

Membrane Samples	WU (wt%)	$MP10^8$ (cm^2/sec)	IEC (mmol/g)	CP ($/g)
SPEEK/PTCMA (0)	15.00	28.75	1.64	4.57
SPEEK/PTCMA (10)	21.74	16.50	1.75	4.30
SPEEK/PTCMA (20)	16.67	5.15	1.76	4.02
SPEEK/PTCMA (30)	14.29	4.82	1.80	3.75
SPEEK/PTCMA (40)	14.81	0.28	1.97	3.48
SPEEK/PTCMA (50)	11.11	0.56	2.20	3.20

WU: Water Uptake; MP: Methanol Permeability; IEC: Ion Exchange Capacity; CP: Cost of Polymer

Results and Discussion

Table II. Relative Importance of Performance Goals Using Modified Digital Logic Method

Goals	Number of Possible Decisions						Positive Decision	Weighting Factor(α)
	1	2	3	4	5	6		
WU	2	2	3				7	0.29
MP	2			1	2		5	0.21
IEC		2		3		3	8	0.33
CP			1		2	1	4	0.17

WU: Water Uptake; MP: Methanol Permeability; IEC: Ion Exchange Capacity; CP: Cost of Polymer

Results and Discussion

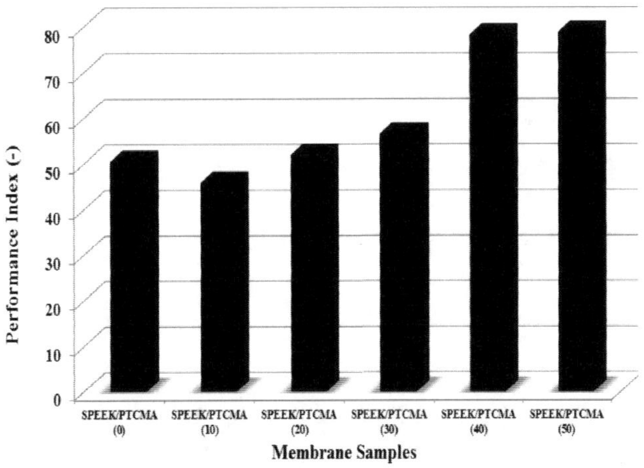

Figure 1 Performance of membrane materials based on overall performance index

Conclusion and Recommendations

- The use of multi-objective optimization method for optimizing PEM material selection was demonstrated
- SPEEK/ PTCMA (40) and SPEEK/ PTCMA (50) were the best two composites
- Overall performance indices were 78 and 79, respectively
- Outcomes is an indication of the suitability of multi-objective optimization approach to optimizing PEM material selection

Conclusion and Recommendations

- Studies of more PEM performance properties
- Upper and lower limits of performance properties (Constrained optimization)
- Model based approach and fuel cell performance

References

[1] H.-L. Wu, C.-C. M. Ma, F.-Y. Liu, C.-Y. Chen, S.-J. Lee, and C.-L. Chiang, Preparation and Characterization of Poly(Ether Sulfone)/Sulfonated Poly(Ether Ether Ketone) Blend Membranes, Eur Polym J. **42**(7) (2006):1688-1695.

[2] Drioli E., Regina A., Casciola M., Oliveti A., Trotta F., and M. T., Sulfonated Peek-Wc Membranes for Possible Fuel Cell Applications, . J M embr Sci **228** (2004):139–48.

[3] Carrette L, Friedrich K. A., and S. U., Fuel Cells-Fundamentals and Applications, Fuel Cells **1**(2000):5-39.

[4] S. Ren, C. Li, X. Zhao, Z. Wu, S. Wang, G. Sun, Q. Xin, and X. Yang, Surface Modification of Sulfonated Poly(Ether Ether Ketone) Membranes Using Nafion Solution for Direct Methanol Fuel Cells, Journal of Membrane Science. **247**(1–2) (2005):59-63.

[5] S. Zhong, X. Cui, T. Fu, and H. Na, Modification of Sulfonated Poly(Ether Ether Ketone) Proton Exchange Membrane for Reducing Methanol Crossover, J Power Sources. **180**(1) (2008):23-28.

[6] F. Lufrano, V. Baglio, P. Staiti, A. S. Arico', and V. Antonucci, Polymer Electrolytes Based on Sulfonated Polysulfone for Direct Methanol Fuel Cells, J Power Sources. **179**(1) (2008):34-41.

[7] A. Wolz, S. Zils, M. Michel, and C. Roth, Structured Multilayered Electrodes of Proton/Electron Conducting Polymer for Polymer Electrolyte Membrane Fuel Cells Assembled by Spray Coating, J Power Sources. **195**(24) (2010):8162-8167.

References

[8] F. Lufrano, V. Baglio, P. Staiti, A. Stassi, A. S. Aricò, and V. Antonucci, Investigation of Sulfonated Polysulfone Membranes as Electrolyte in a Passive-Mode Direct Methanol Fuel Cell Mini-Stack, J Power Sources. **195**(23) (2010):7727-7733.

[9] X. Li, C. Liu, D. Xu, C. Zhao, Z. Wang, G. Zhang, H. Na, and W. Xing, Preparation and Properties of Sulfonated Poly(Ether Ether Ketone)S (Speek)/Polypyrrole Composite Membranes for Direct Methanol Fuel Cells, J Power Sources. **162**(1) (2006):1-8.

[10] Y.-S. Ye, J. Rick, and B.-J. Hwang, Water Soluble Polymers as Proton Exchange Membranes for Fuel Cells, Polymers. **4**(2) (2012):913-963.

[11] Q. Li and J. O. Jensen, Membranes for High Temperature Pemfc Based on Acid-Doped Polybenzimidazoles, in Membranes for Energy Conversion. 2008, Wiley-VCH Verlag GmbH & Co. KGaA. p. 61-96.

[12] S. Mukherjee, A. Bates, S. C. Lee, D.-H. Lee, and S. Park, A Review of Application of Cnts in Pem Fuel Cells, International Journal of Green Energy. (2014):null-null.

[13] Q. Zeng and A. Yu, Prediction of the Mechanical Properties of Nanocomposites, in Optimization of Polymer Nanocomposite Properties, M. V., Editor. 2010, WILEY-VCH Verlag GmbH & Co KGaA: Weinhein. p. 301 - 331.

KFUPM

© King Fahd University of Petroleum & Minerals | Dhahran, 31261 | Saudi Arabia | +966 (3) 860 0000

References

[14] A. Jahan, M. Y. Ismail, S. M. Sapuan, and F. Mustapha, Material Screening and Choosing Methods – a Review, Materials & Design. **31**(2) (2010):696-705.

[15] B. Dehghan-Manshadi, H. Mahmudi, A. Abedian, and R. Mahmudi, A Novel Method for Materials Selection in Mechanical Design: Combination of Non-Linear Normalization and a Modified Digital Logic Method, Materials & Design. **28**(1) (2007):8-15.

KFUPM

Acknowledgement

- National Science, Technology and Innovation Program (NSTIP) of KACST (under project number 10-ENE1374-04)
- Center of Petroleum & Minearls
- Center of Excellence in Renewable Energy

KFUPM

© King Fahd University of Petroleum & Minerals | Dhahran, 31261 | Saudi Arabia | +966 (3) 860 0000

جامعة الملك فهد للبترول والمعادن
King Fahd University of Petroleum & Minerals
Thank You!
© King Fahd University of Petroleum & Minerals | Dhahran, 31261 | Saudi Arabia | +966 (3) 860 0000

422190 Predictive Modeling Insights for Active Three-Phase Boundary Sites in Solid Oxide Fuel Cell Infiltrated Electrodes

Abstract:

A Solid Oxide Fuel Cell (SOFC) is an electrochemical device that converts chemical energy into electrical energy by directly and efficiently oxidizing hydrocarbon fuels at high temperature. Fuel oxidation and oxygen reduction reactions occur in the electrodes. For an electrode to be functional the following four properties must present: 1) porosity for mass transport, 2) catalytic activity for the reaction, 3) oxygen ion conductivity, and 4) electronic conductivity. Infiltrated electrodes, where electronic conductor/catalyst particles are infiltrated into an oxygen ion conducting porous scaffold, can result in outstanding electrochemical performance. In addition, infiltration allows one to calcine the infiltrated particles at much lower temperatures than the scaffold sintering temperature. The benefits conferred include less electronic conducting material to form an interconnected conductive pathway, reduced dimensional instabilities within the fuel cell, and greater control over the infiltrate morphology. While the infiltration method provides great flexibility in composite composition and structure, the influence of the controllable parameters are not well understood. Furthermore, fabrication and testing of anode composites representing the full range and number of working parameters is time consuming. Thus, a model to guide experiments is desired.

In previous work, we have demonstrated the ability of our mechanistic model to predict effective conductivity of the fuel cell electrode as a function of important processing variables such as scaffold particle and pore size, infiltrate size and loading, porosity, and initial scaffold film formulation. In this talk, we will highlight our recent work, building on this model, to predict the density of active three phase boundary (TPB) sites in the electrode as a function of the same processing variables. These TPB sites, the location where the electrochemical reaction occurs, are located at the intersection of (1) void space, allowing for the presence of gas-phase reactants, (2) electronically conducting/catalytically active particles and (3) oxygen ion conducting backbone. The model itself contains no adjustable parameters and is based upon the underlying physics of the entire fabrication process, which includes the formation of a presintered film, evaporation of the solvent in that film, burning off of the pore formers in that film, sintering of the ceramic scaffold and infiltration of conducting particles. In addition to reporting the resulting predicted TPB densities and their relation to corresponding electronic conductivities, we identify the key underlying limiting structural feature for TPB density, interfacial surface areas of each pair of phases. Further, we provide insight into the way these limiting features as well as the other modeling results suggest areas for expanding the experimental research design space. Finally, we will discuss the opportunities to use this type of modeling to incorporate other SOFC properties leading to overall design optimization.

405597 Preparation and Performance of Ni-Co/SiO2 Catalyst for Catalytic Decomposition of Methane to Cox Free Hydrogen

Abstract:

A prepared and commercial Ni catalyst was employed in a thermo-catalytic decomposition of methane to produce COx-free hydrogen. Ni catalysts supported on silica was prepared by wet impregnation method are studied to produce pure hydrogen and carbon fibers in a fixed-bed quartz reactor at 500-600 °C and at atmospheric pressure.

It was observed that the prepared10% Ni-Co/ Al_2O_3 catalyst had a good activity in methane decomposition reaction with a high activity at the initial times of reaction and decreased with time on stream, and finally reached a constant value.

The experimental results of methane conversion versus temperature and time on stream were subjected to the integral method of analysis to predict the rate law of the catalytic decomposition. The reaction order for methane decomposition over the two types of Ni catalysts is approximately 2. An activation energy of 148 kJ/mol for the prepared 10% Ni-Co/ $Al_2O_3$and 100.7 kJ/mol for the commercial 12% Ni-SiO_2/Al_2O_3catalysts.

435130 Thermodynamic Studies of a Novel Water Splitting Thermochemical Cycle

Abstract:

A theoretical study of the constitutive reactions from a novel water splitting thermochemical cycle is presented, which aims to find their respective operating conditions as part of the cycle. The cycle is categorized as a three-step cycle that is based on the thermal decomposition of sodium carbonate Na2CO3 at temperatures higher than 1131.25 K (melting point of sodium carbonate). The chemical species involved in the cycle are: water H2O, oxygen O2, hydrogen H2, sodium Na, sodium carbonate Na2CO3, sodium hydroxide NaOH, and carbon dioxide CO2. The chemical reactions of the cycle are:

1) $2Na_2CO_3 \leftrightarrow 4Na + 2CO_2 + O_2$

2) $4Na + 4H_2O \leftrightarrow 4NaOH + 2H_2$

3) $4NaOH + 2CO_2 \leftrightarrow 2Na_2CO_3 + 2H_2O$

with lead to the overall water decomposition reaction:

4) $2H_2O \leftrightarrow 2H_2 + O_2$

Reaction 1) is endothermic, and thus, heat needs to be provided in order for the decomposition of sodium carbonate to be carried out, typically at temperatures higher than its melting point. Reactions 2) and 3) are exothermic.

In this work, each individual reaction involved in the cycle is analyzed thermodynamically in order to find the temperature, pressure, sweeping gases, and flowrate/dilution conditions ranges for which each reaction can be carried out. This is done through the Gibbs minimization method which allows to find condition ranges due to its flexibility on changing parameters. This work provides fundamental information regarding the cycle's reactions species at high temperature and low pressure, not currently available in the literature. Additionally, the results from this work pave the way for the design of potential flowsheets that can be subjected to optimization studies and even coupled with a Rankine power cycle in order to co-generate electricity.

431523 Economic and Environmental Analysis of Liquid Fuel Production from Coal and Coal/Biomass Mixtures

Abstract:

Fischer-Tropsch synthesis (FTS) is the preferred basis for the indirect liquefaction of coal and biomass for the production of liquid transportation fuels. Conventional FTS is highly unselective and produces lots of solid wax that requires additional equipment to convert the wax through a hydrogen intense process before it can be fractionated into separate liquid transportation fuels. This significantly increases both capital and operating costs for both treating the waxes and the generation of the hydrogen.

Based upon the results of experiments conducted under DOE/NETL sponsored cooperative agreement to demonstrate a compact, pressurized, high temperature reformer and a highly selective cobalt-zeolite hybrid Fischer-Tropsch (FT) catalyst a techno-economic analysis (TEA) and life-cycle analysis (LCA) was conducted to show the effects of these new technologies when compared to conventional indirect liquefaction coal-to-liquid (CTL).

The TEA showed that the improved technologies were able to reduce the capital cost of the indirect CTL process by; eliminating the need for several parasitic gas cleaning steps, eliminating the need for FTS product upgrading, and reducing the need for hydrogen generation. These cost reductions improved the economic competitiveness of the CTL process. The LCA showed that the addition of a moderate amount of biomass reduced the overall greenhouse gas emissions associated with the liquid fuels below that of a comparable petroleum derived fuel. The addition of CO_2 capture technology can further reduce the greenhouse gas emissions, however, the associated costs can significantly alter the cost competitiveness.

412454 Bioconversion of Natural Gas for the Production of Isobutanol Using a Methanotroph

Abstract:

Bioconversion of natural gas to iso-butanol : Engineering considerations for commercialization

The low cost and abundant supply of natural gas have been motivators for the development of many technologies that convert this low cost carbon feedstock into compounds of greater value. Natural gas bioconversion is an example of a developmental technology which exploits a methanotroph, a methane consuming bacteria, by applying synthetic biology to program it to produce higher value materials of interest, such as biofuels or terpenes. Intrexon Energy Partners was formed as a joint venture to commercialize the production of isobutanol from natural gas using a methanotroph. Currently, the production of both isobutanol and farnesene has been demonstrated in lab scale fermentation. However, to successfully bring this technology to commercial scale, consideration must be given not only to optimizing the metabolic pathway engineered into the methanotroph, but to key issues in the fermentor design and downstream processing. Since the substrate is a gas, the design of the fermentor must be optimized for effective mass transfer of the natural gas to the liquid phase for increased bioavailability. Isobutanol is toxic to the organism above a certain level, so the process must also be designed to continuously remove isobutanol in order to maintain the fermentor concentration below the toxic threshold. Downstream processing must take into consideration purification of the isobutanol and removal of fermentation byproducts and contaminants. Process selection must weigh the efficiency and selectivity of technology options, while at the same time utilize as much mature technology as possible to minimize risk. The project team has performed numerous design reviews known as Front End Loading (FEL) to develop process criteria and scenarios that will be applied to the first demonstration plant.

417457 Microbial Conversion of Natural Gas to Biofuel Using a Naturally Occurring Methanotroph

Abstract:

Methane, the second most prominent greenhouse gas, comes from multiple sources including the production of and refining of petroleum, domestic livestock, and landfills, and accounts for approximately 9% of greenhouse gas emissions in the United States according to the EPA.[1] This research focuses on the conversion of methane to methanol, which can be used in the process of biofuel production. The bacterium *Methylosinus trichosporium* OB3b, a naturally occurring methanotroph, is being evaluated to promote this conversion.

M. trichosporium OB3b cells, when immobilized in an alginate gel solution, maintain their metabolic activity. This technique was used to immobilize the cells in alginate beads and in a thin film to investigate the kinetic properties of the cells in various reactor settings. The alginate beads were placed in a packed column with liquid up-flow capabilities for long term experiments to determine the viability of the culture. Immobilizing the culture in alginate allows it to be easily used and evaluated in a packed bed column, while maintaining mass transfer properties. A methane and oxygen saturated solution was passed through the packed column where methane was oxidized to methanol by the methane monooxygenase (MMO) enzyme. To stop the further metabolism of methanol, cyclopropanol was used to inhibit the methanol dehydrogenase (MDH) enzyme, resulting in methanol accumulation. Long and short inhibition cycles have been evaluated at hydraulic residence times of 5-10 minutes with the overall goal of long term methanol production. Methanol concentrations of 0.1 mmol/L have been obtained. The culture has sustained lower levels of activity in the packed column for up to two weeks, through cycles of inhibition and methanol accumulation, which supports long term methanol production capabilities.

For the further investigation of the *M. trichosporium* OB3b activity, immobilization techniques and inhibition processes within the alginate gel beads are being analyzed using a microelectrode oxygen probe. The cells are also being immobilized in a thin alginate film on a flat surface metal slide. The alginate film is approximately 1 mm thick and contains a cell concentration of around 10 g/L. This slide is placed in a laminar flow bath with a media solution flowing over the stationary film. A dissolved oxygen microelectrode (10 μm diameter) is being used to evaluate the oxygen profile in the boundary layer above the film and 500 μm into the film. Estimates of the flux of dissolved oxygen (DO) into the film will be used to calculate methane and methanol uptake within the film. This process was repeated with concentrations of 1 and 10 ppm methanol in the laminar flow bath. The profiles are being created and evaluated for the various concentrations of methanol. The immobilized cells on the slides are also being inhibited with cyclopropanol to evaluate the inhibited oxygen uptake for comparisons. A modified system was created in batch mode with a thick layer of alginate beads with immobilized OB3b. A DO profile was measured with no added substrate and then methanol was added at 1 ppm and the process was repeated in triplicates.

Future work includes further analysis of the inhibition process using cyclopropanol, the immobilization techniques, their effects on MMO and MDH enzyme activity, and the oxygen and pH changes in a thin film system. These results are being applied to the design of a thin film reactor currently in preliminary stages of investigation for long term, high rate methanol production.

---

[1] *EPA*. Environmental Protection Agency, 2 July 2014. Web. 9 April 2015.

15AIChE
Annual Meeting, Salt Lake City, UT

TECHNO-ECONOMIC ASSESSMENT OF BIODIESEL PRODUCTION FROM PALM OIL BY SUPERCRITICAL TRANSESTERIFICATION

*Caballero Ashley S[1], García Carlos[1], Daza Laura[1], Cardona Carlos A[1**].*
[1] Procesos Químicos, Catalíticos y Biotecnológicos, , Departamento de Ingeniería Química, Facultad de Ingeniería y Arquitectura, Universidad Nacional de Colombia sede Manizales, Manizales, Colombia.
**e-mail:ccardonaal@unal.edu.co*

1

15AIChE
Annual Meeting, Salt Lake City, UT

Objetive | Methodology | Results | Conclusion | Acknowledgements | References

Introduction

BIODIESEL

Obtaining sources

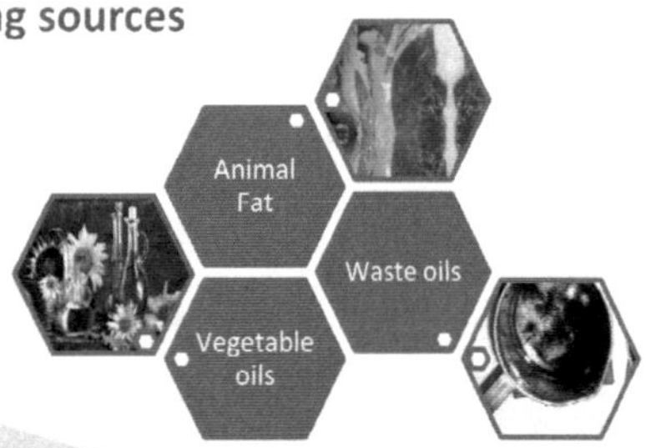

$$Triglycerides + Alcohol \leftrightarrow Biodiesel + Glycerin$$

2

Transesterification reaction

$$Triglycerides + Alcohol \leftrightarrow Biodiesel + Glycerin$$

Basic catalyst

- **Homogeneous:** Catalysts that can be used on a commercial scale (KOH, NaOH, CH 3 ONa) fast acting at moderate condition.
- **Heterogeneous:** MgO, CaO, $Na/NaOH/Al_2O_3$ which they can be reused presenting the disadvantage however of requiring high pressure and temperature.

Acid catalyst.

- **Homogeneous:** such as H_2SO_4, H_3PO_4. No soap formation, but have a low reaction rate and difficulties in purification operation.
- **Heterogeneous:** as zeolites, sulfonic resins. They can be reused, simple purification, present problems of diffusivity.

Enzyme catalyst

- Novozym-435, type Burkholderia cepacia lipase, Candida antarctica, P. fluorescens, Mucos miehei. Facilitate separation steps and reuse avoid soap formation. They have high costs.

Vegetable oil sources in Colombia

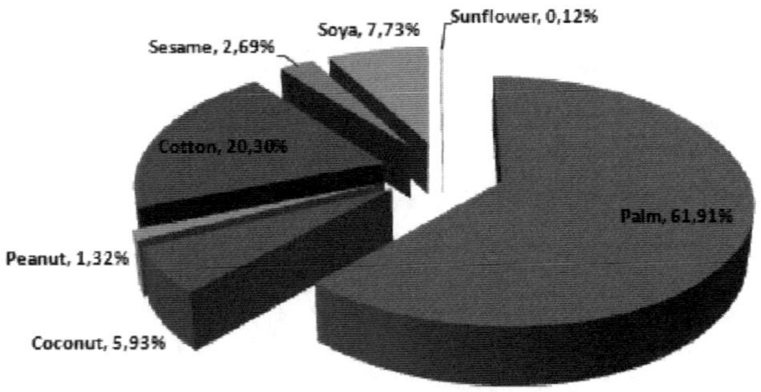

Figure 1. Distribution of oil cultivated area in Colombia. "BIODIESEL" Cardona C., Orrego E., & Gutiérrez L.

Palm oil (*Elaeis guineensis*)

USES OF OIL PALM

- Food industry
- Chemestry indutstry
- Cosmetic
- Animal feed
- Biofuel

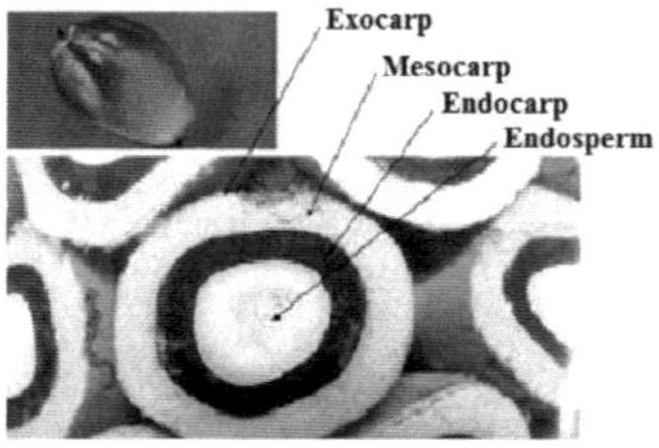

Supercritical Fluid

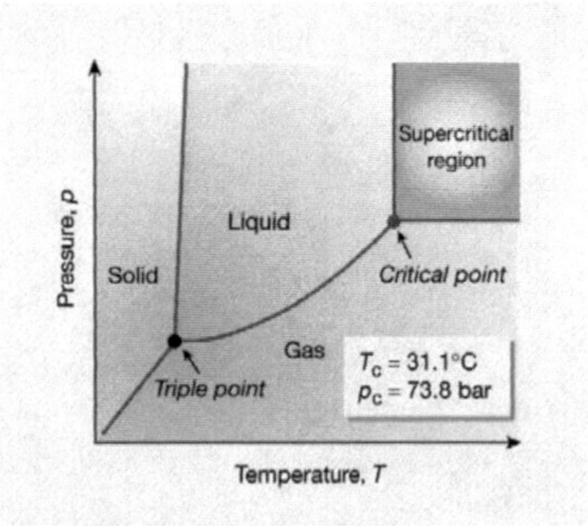

Figure 2. Critical properties of CO_2 "Introducción a las operaciones de separación no convencionales" Cardona C.

Advantages of CO2 as a supercritical fluid.

- It is not toxic.
- It is not explosive.
- It is chemically inert.
- It is not corrosive.
- It is easily recycled.
- Available high purity.

15AIChE

Objetive Methodology Results Conclusion **Acknowledgements** References

Introduction

Supercritical Processes Catalyzed By Enzymes

Lipases

- Triacylglycerol hydrolase.
- Enzymes capable of hydrolyzing, synthesizing glycerides of long chain fatty acids.
- It requires minimal availability of water for these enzymes reach their catalytic function.

Supercritical Processes

- Lipases in supercritical processes increase the yield, selectivity, conversion and reaction rate
- Moderate temperatures to ensure stability of lipases.

7

15AIChE

Introduction Methodology Results Conclusion Acknowledgements References

Objetive

Study the biodiesel in supercritical conditions from palm oil using *Candida rugosa.*

8

15AIChE

Introduction Objetive **Methodology** Results Conclusion Acknowledgements References

1. EXPERIMENTAL PROCEDURE

- *Characterization of raw material*
- *Enzyme activation*
- *Production of biodiesel*
- *Characterization of Biodiesel*

2. SIMULATION

- *Balances de materia y energía*
- *Kinetic Data*
- *Economic analysis*

9

15AIChE

Introduction Objetive **Methodology** Results Conclusion Acknowledgements References

Characterization of raw material

Density
- ASTM D4052

Viscosity (40°C)
- ASTM D445

Humidity content

Acidity Index
- ASTM D664

10

Enzymatic activation

Lipase submersion tert butanol (1 hour) → Vacuum filter → Add lipases palm oil (50 minutes)

15AIChE

Introduction Objetive **Methodology** Results Conclusion Acknowledgements References

Production of biodiesel

Table 1. Process specifications

Pressure	180 bar
Temperature	50 °C
Alcohol	Ethanol
Time	4 hours
Catalyst	Lipases (*Candida rugosa*)

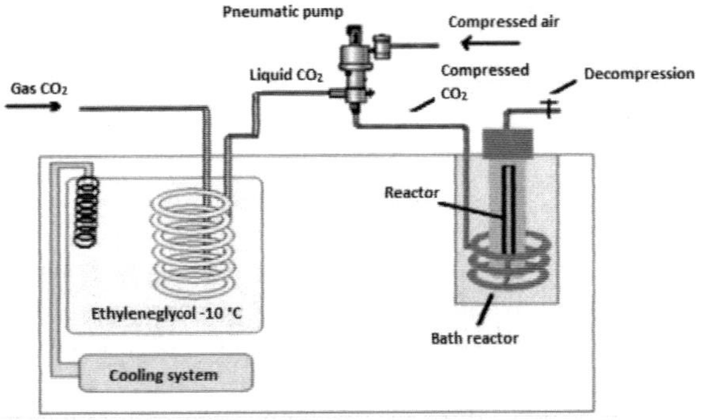

Figure 3. Diagram supercritical equipment.
"Comparación de procesos para producir biodiesel en condiciones supercríticas" Cúbides D.

15AIChE

Introduction Objetive **Methodology** Results Conclusion Acknowledgements References

Product characterization

Density
- ASTM D4052

Viscosity
- ASTM D445

Humidity content

Acidity Index
- ASTM D664

Corrosion of copper foil
- ASTM D130-12

Iodine index
- BS EN 14111:2003

13

15AIChE

Introduction Objetive **Methodology** Results Conclusion Acknowledgements References

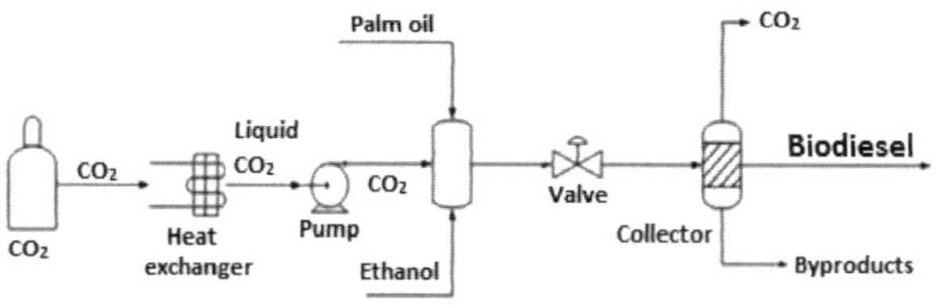

Figure 4. Flow diagram of production of biodiesel in the supercritical reactor

Techno-economic assessment

Material and energy balances

Aspen Plus V8.2 (ASPEN TECHNOLOGY INC)

Kinetic data.

Economic analysis

Calculated with the use of Aspen Process Economic Analyzer software V8.2 (ASPEN TECHNOLOGY INC), applying the conditions of Colombia.

14

15AIChE

Introduction Objetive Results Conclusion Acknowledgements References

Methodology

Feed ratio of 1: 5 oil-alcohol

Table 2.
Palm oil composition [8]

Compound	% Mass of oil fed
Tripalmitin	47.14
Trilinoein	37.52
Triolein	11.52
palmitic acid	3.82

Kinetic data

$$r = k \times e^{\left(-Ea/_{RT}\right)}$$

Table 3.
Kinetic data [8]

	Value	Units
k	0.007	-----
Ea	38.483	KJ/Kmol

15

15AIChE

Introduction Objetive Methodology Conclusion Acknowledgements References

Results

Figure 5. Biodiesel.

Table 4.
Physico-chemical characteristics of palm oil

Property	Value	Units
Density (15°C)	0.871	g/ml
Viscosity (40°C)	37.3	cp
Humidity content	7.90	%
Acidity Index	5.83	%

Table 5.
Yield enzyme production of biodiesel in supercritical conditions (180 bar and 50 ° C for this job)

	Experimental value this work	Experimental literature value.	Value simulation This work	Bibliographical value (with T higher than 70 ° C)
Process Yield (%)	70	74 - 97 [12]	90	< 99 [12]

16

Introduction Objetive Methodology **Results** Conclusion Acknowledgements References

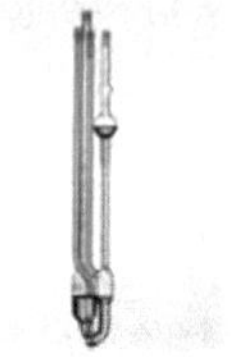

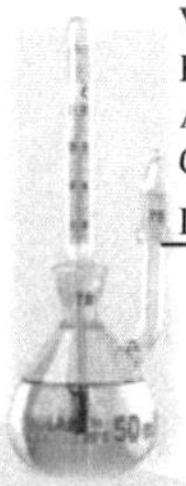

Table 6.
Physico-chemical characteristics of biodiesel

Property	Experimental value	Permissible value	Units
Density (15 °C)	0.880	0.860-0.900	g/ml
Viscosity (40 °C)	3.8	1.9-6.0	mm^2/s
Humidity content	375	máx 500	mg/kg
Acidity Index	0.1	máx 0.5	mg KOH/g
Corrosion of copper foil	1	1	
Iodine index	60	máx 120	g yodo/100g

17

Introduction Objetive Methodology **Results** Conclusion Acknowledgements References

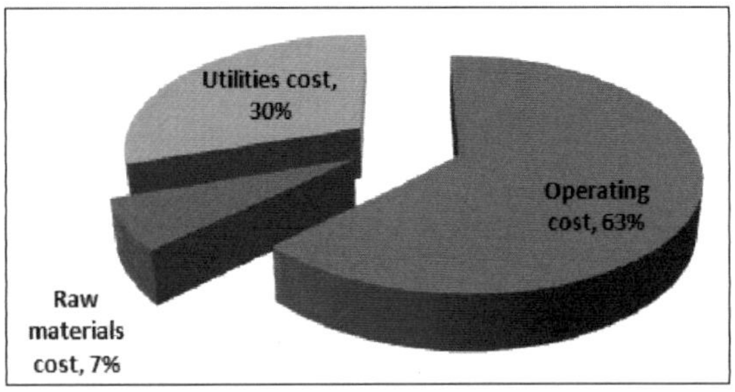

Figure 6. Percentage of production cost of biodiesel.

Table 7.
Production of biodiesel

Production cost of biodiesel (USD / L))	0.41
Process gain (%)	32

18

15AIChE

Introduction Objetive Methodology Results Acknowledgements References

Conclusions

▪The decrease and residence or reaction time is so high that biodiesel increases productivity and capital costs and amortization are compensated in the plant with supercritical CO2.

▪The experimental evaluation demonstrated that it is possible to obtain high-quality biodiesel with pressures close to 180 atm and temperatures close to 50 ° C using lipases.

▪The simulation demonstrated economic feasibility of biodiesel production scales of 2800kg / h. Is worth noting that the results can be improved markedly if an energy recovery cycle in the supercritical scheme is involved.

19

15AIChE

Introduction Objetive Methodology Results Conclusion References

Acknowledgements

The authors wish to express their gratitude to:

❖23132 Project Hermes.

❖Universidad Nacional de Colombia for supporting this research.

❖Instituto de Biotecnología y Agroindustria.

20

15AIChE

Introduction Objetive Methodology Results Conclusion Acknowledgements

References

- Palmas, "PALMA ACEITERA CULTIVO y AGROINDUSTRIA: Oportunidad para el desarrollo ordenado de la Amazonía," 2012.
- U. Del and A. D. E. Palma, "ACEITE DE PALMA : USOS , ORIGENES E IMPACTOS," 2007.
- R. Davarnejad, K. M. Kassim, a. Zainal, and S. a. Sata, "Supercritical fluid extraction of β-carotene from crude palm oil using CO2," *J. Food Eng.*, vol. 89, pp. 472–478, 2008.
- Superintendencia de Industria y Comercio, "Agroindustria de la Palma Africana: Diagnostico de Libre Competencia," pp. 1–9, 2011.
- T. Muppaneni, H. K. Reddy, S. Ponnusamy, P. D. Patil, Y. Sun, P. Dailey, and S. Deng, "Optimization of biodiesel production from palm oil under supercritical ethanol conditions using hexane as co-solvent: A response surface methodology approach," *Fuel*, vol. 107, pp. 633–640, 2013.
- K. T. Tan, K. T. Lee, and a. R. Mohamed, "Potential of waste palm cooking oil for catalyst-free biodiesel production," *Energy*, vol. 36, no. 4, pp. 2085–2088, 2011.
- S. a. Biktashev, R. a. Usmanov, R. R. Gabitov, R. a. Gazizov, F. M. Gumerov, F. R. Gabitov, I. M. Abdulagatov, R. S. Yarullin, and I. a. Yakushev, "Transesterification of rapeseed and palm oils in supercritical methanol and ethanol," *Biomass and Bioenergy*, vol. 35, no. 7, pp. 2999–3011, 2011
- C. A. Cubides, Diana; Cardona, "Comparacion de procesos para producir biodiesel en condiciones supercriticas," 2009.
- M. Aarthy, P. Saravanan, M. K. Gowthaman, C. Rose, and N. R. Kamini, "Enzymatic transesterification for production of biodiesel using yeast lipases: An overview," *Chem. Eng. Res. Des.*, vol. 92, no. 8, pp. 1591–1601, 2014.
- Y. Li, W. Du, and D. Liu, "Efficient biodiesel production from phospholipids-containing oil: Synchronous catalysis with phospholipase and lipase," *Biochem. Eng. J.*, vol. 94, pp. 45–49, 2015.
- Y. Li, W. Du, and D. Liu, "Free lipase-catalyzed biodiesel production from phospholipids-containing oils," *Biomass and Bioenergy*, vol. 71, pp. 162–169, 2014

21

15AIChE

Introduction Objetive Methodology Results Conclusion Acknowledgements References

Thanks for your attention

*Caballero Ashley S[1], García Carlos[1], Daza Laura[1], Cardona Carlos A[1**].*

[1] Procesos Químicos, Catalíticos y Biotecnológicos, , Departamento de Ingeniería Química, Facultad de Ingeniería y Arquitectura, Universidad Nacional de Colombia sede Manizales, Manizales, Colombia.
(+57) (6) 879400 ext 55354.
[**] Corresponding author: ccardonaal@unal.edu.co

22

434389 A Novel Foam Photobioreactor for Algal Wastewater Treatment, CO2 Capture and Biofuel Production

Abstract:

This study investigated a novel algal foam photobioreactor to support sustainable wastewater treatment, CO_2 Capture and biofuel production. The novel algal foam bioreactor relied on the actively growing microalgae in tiny aqueous foam bubbles (~ 2 mm size, 150 μm liquid film thickness) which provided high mass and light transfer in the bioreactor. The microalgae in the foam photobioreactor were isolated from the aquaponic system at University of Hawaii operated with high nitrogen and phosphorus concentration. The synthetic and real sewage wastewater were treated by the algal bioreactor when the microalgae were cultivated in three growth modes (autotrophic, mixotrophic and heterotrophic cultivation) over the long term period. Three growth modes had the same nutrients (total nitrogen and phosphorus) and supplements from the synthetic and real wastewater except CO_2 for the autotrophic, CO_2/BOD for the mixotrophic, and BOD for the heterotrophic culture. The results showed that the mixotrophic culture of algae showed highest cell growth and chlorophyll content among three growth modes. The mixotrophic culture of algae also exhibited its removal of carbon (organic contaminant), nitrogen and phosphorus in the wastewater which were much higher than the autotrophic and heterotrophic culture of algae. Analysis of algal composition (lipid, carbohydrate, protein) showed high potential for production of bio-oil and biochar while it would be also used for biodiesel production. In addition to algal wastewater treatment, the algal foam photobioreactor led to high CO_2 removal at short gas contact time (i. e., 30-70% CO_2 removal at gas contact time of 60 s) of CO_2 as a potential CO_2 capture. In overall, these results supported that the novel algal foam photobioreactor integrated with wastewater treatment, CO_2 capture and biofuel production algal treatment of wastewater would be developed as a cost-effective sustainable bioprocess.

435183 Lipid Production from Lignocellulosic Biomass:Efficient Strategies for Prtreatment Induced Inhibitors

Abstract:

Lipid production from lignocellulosic biomass: Efficient strategies for pretreatment-induced inhibitors

Jungeun Lee[1], Yadhu N. Guragain[1], Krishna P.Bastola[1], Praveen V.Vadlani[1,2]

Bioprocessing and Renewable Energy Laboratory, Department of Grains Science and Industry[1], Department of Chemical Engineering[2], Kansas State University, Manhattan, Kansas, USA

Oleaginous yeast containing triacylglycerides can be effectively produced from lignocellulosic biomass. For utilization of lignocellulosic biomass, pretreatment methods are employed to obtain sugars via effective depolymerization. However, since harsher methods are employed during pretreatment, toxic compounds are generated during monomer sugar production. This study focuses on innovative strategies to alleviate toxic compounds, which improve lipid production within oleaginous yeast. Initially, absorption was adopted to remove toxic compounds from monomer sugar solution. Activated charcoal and resins were tested to remove toxic compounds from sorghum hydrolysates. Activated charcoal efficiently removed hydrophobic compounds such as hydroxymethylfurfural (HMF) and furfural. However, acetic acid was not removed, and 3% of sugar loss was detected. Whereas, resin mixture (7:3 ratio of cation and anion exchangers) removed all types of inhibitors without sugar loss. This result showed that detoxification column with resin mixture was an efficient method for detoxifying hydrolysates, and it is anticipated that this method will improve lipid production during fermentation. Other approach to handle toxic compounds was selection of robust oleaginous yeast for lipid production. In our study, three types of oleaginous yeast (*Trichosporon oleaginous, Lipomyces starkeyi* and *Cryptococcus albidus*) with high tolerance to toxic compounds were evaluated using sorghum hydrolysates. *T. oleaginosus* accumulated about 40% of lipids, and produced total 6.2g/L of lipids at 120h fermentation. *L. starkeyi* and *C.albidus* accumulated about 40% of lipids, and produced total 4g/L of lipids. Our experiments confirmed that T. oleaginosus is the appropriate microbial system for effective production of lipids from sorghum hydrolysates.

434403 Process Simulation of Biogas to Renewable Liquid Fuels

Abstract:

Biogas is a product of anaerobic digestion of waste organic sources such as animal manure, food wastes and pretreated lignocellulosic biomass. Biogas is primarily composed of mixtures of 50–75% CH_4, 25–50% CO_2, 0–10% N_2, 0–5%O_2, 1000-2000 ppm H_2S. Biogas-to-Liquid (BTL) technology is relatively mature to produce wide range of fuels and chemicals via Fischer Tropsch synthesis. In this study, a process based simulation model was developed using an Aspen Plus® simulation tool. The primary feedstock is a raw biogas from liquid state or solid state anaerobic digestion plant, which was upgraded to biomethane (>96% methane content) using a high pressure water scrubbing (HPWS®) technology with a 96% upgrading efficiency. Biomethane, as similar to natural gas is converted to synthetic gas using both partial oxidation and Ni catalyst driven steam reforming in order to obtain the approximate H_2/CO ratio of 2 necessary for the FT process. A low temperature slurry-bed FT reactor was used for the conversion of syngas at a temperature of 220-240°C in the presence of cobalt catalyst/support to generate a wide range of hydrocarbon chain products. Upgrading of the hydrocarbon products using distillation and hydrocracking processes yields diesel, LPG and gasoline fuels. The detailed mass and energy balance for each unit operation was conducted and presented along with products yield and distribution. Finally, the techno-economic model was developed to estimate the cost of drop-in fuels from biogas.

(500b) Production of High Surface Area Activated Carbon from Fast Pyrolysis Biomass Char By Chemical Methods

Luo, Y.,
Guda, V. K., Mississippi State Univeristy
Steele, P., Mississippi State University
Originally presented on: 11/11/2015 8:52:00 - 9:14:00

Production of high surface area activated carbon from fast pyrolysis biomass char by chemical methods

Yan Luo, Vamshi Krishna Guda, Philip H. Steele

Department of Sustainable Bioproducts, Mississippi State University, Mississippi State, MS 39762, United States

Dave C. Swalm School of Chemical Engineering, Mississippi State University, Mississippi State, MS 39762, United States

Currently, activated carbon is produced mostly from coal with some produced from petroleum coking. Development of bioproducts from non-fossil fuel resources is now a national focus area. Biomass char has a potential to produce activated carbon as the same qualities that current by attained from petroleum. Fast pyrolysis char has a surface area approximately 200 m^2/g without any treatments.

In this study, four types of fast pyrolysis char, including pine wood, bamboo, switch-grass and giant miscanthus char were applied to prepare activated carbon by chemical methods. KOH, H_2O_2 and H_3PO_4 were applied to the pyrolysis char to attempt to increase surface area. The weight ratio of the char to chemical activator was 1:1. The surface areas of all activated carbon prepared from the four types of fast pyrolysis char were increased dramatically at an activation temperature of 750 °C for 1 h at a heating rate of 5 °C/min. Activated carbon with the highest surface area was prepared from KOH-treated fast pyrolysis char. Following KOH treatment, the surface area of pine wood, bamboo, switch-grass and giant miscanthus char were increased to 1124.4, 1200.0, 1432.7 and 1435.8 m^2/g, respectively.

404874 Production of Synthetic Fuel from Anaerobic Digester Gas

Abstract:

The waste generated from feed lots, dairy farms, meat processing, etc can be converted to a mixture of methane and carbon dioxide using anaerobic digestion. The methane can be converted to synthesis gas and subsequently converted to synthetic fuel using Fischer Tropsch technology.

Ceramatec is working with a number of small companies to determine the feasibility of construction of modular digester based Gas-to-Liquids (GTL) plants for the conversion of waste to liquid fuels. Ceramatec is currently operating a small, ¼ BPD Gas-to-Liquids pre-pilot plant at its Salt Lake City, Utah facility. The existing laboratory system has been run with a variety of catalysts and a variety of conditions. Based on the test results, a modular GTL plant that is road transportable has been designed and fabrication cost estimated on a per barrel-per-day (BPD) of capacity. The paper will describe the test results that have been obtained to date and discuss the amount of waste required for a GTL plant based on anaerobic digester gas. The paper will further discuss computer simulation of animal based waste systems both in the United States and overseas.

434486 Catalytic Fast Pyrolysis of Cellulose to Aromatics with Nanocrystalline ZSM-5: Role of Extra-Framework Aluminum

Abstract:

The pyrolysis of lignocellulosic biomass represents a simple, cheap, and efficient approach to produce bio-based fuels and chemicals from renewable feedstocks. In this process, solid biomass is heated to high temperature (500 – 700 °C) in order to thermochemically convert it to light gases (CO, CO_2), solid char, and organic vapours, which can be further condensed to obtain the desired liquid bio-oil. However, the high oxygen content of the bio-oil makes it unsuitable for direct use. Additional processes that involve one or several heterogeneous catalysts are required to decrease the oxygen concentration from approximately 45% to less than 7% and achieve stable blends with petroleum that allow refining. Integrated approaches where the catalyst is directly mixed with the biomass are particularly appealing as pyrolysis and deoxygenation occur simultaneously in the same reactor. Notably, catalytic fast pyrolysis (CFP) using ZSM-5 zeolite as a catalyst produces benzene, toluene, xylene, and naphthalene, which can be used as building blocks by the petrochemical industry or further converted to gasoline-range hydrocarbons using hydrogenation processes already employed in refining. Previous investigations using ZSM-5 led to aromatics yields of 27-30% and no clear correlation with mesoporosity, crystal size, or crystallinity was observed, most probably because of the complexity of the system. Here, we synthesized a series of nanocrystalline ZSM-5 samples and characterized them systematically using nitrogen physisorption, XRD (with internal standard), FE-SEM, ^{27}Al MAS-NMR, NH_3-TPD, and FTIR of adsorbed pyridine. Extra-framework aluminum species were found to be detrimental and favor coke formation. The yield to desired products was increased by up to 135% by controlling crystallinity and aluminum insertion in the zeolitic framework.

405065 The Application of Two-Step Linear Temperature Program to Thermal Analysis for Monitoring the Lipid Induction of Nostoc Sp. KNUA003 in Large Scale Cultivation

Abstract:

Recently, microalgae was considered as a renewable energy for fuel production because its production is nonseasonal and may take place on nonarable land. Despite all of these advantages, microalgal oil production is significantly affected by environmental factors. Furthermore, the large variability remains an important problem in measurement of algae productivity and compositional analysis, especially, the total lipid content. Thus, there is considerable interest in accurate determination of total lipid content during the biotechnological process. For these reason, various high-throughput technologies were suggested for accurate measurement of total lipids contained in the microorganisms, especially oleaginous microalgae. In addition, more advanced technologies were employed to quantify the total lipids of the microalgae without a pretreatment. However, these methods are difficult to measure total lipid content in wet form microalgae obtained from large-scale production. In present study, the thermal analysis performed with two-step linear temeperature program was applied to measure heat evolved in temperature range from 310 to 351 °C of *Nostoc* sp. KNUA003 obtained from large-scale cultivation. And then, we examined the relationship between the heat evolved in 310 to 351 °C (H_E) and total lipid content of the wet *Nostoc* cell cultivated in raceway. As a result, the linear relationship was determined between H_E value and total lipid content of *Nostoc* sp. KNUA003. Particularly, there was a linear relationship of 98% between the H_E value and the total lipid content of the tested microorganism. Based on this relationship, the total lipid content converted from the heat evolved of wet *Nostoc* sp. KNUA003 could be used for monitoring its lipid induction in large-scale cultivation.

419232 Biodiesel Produced from Acidic Vegetable Oils with Zeolite ANA and CAN As Heterogeneous Catalysts

Abstract:

In this study, Zeolite ANA and CAN are adopted to catalyze transesterification of acidic vegetable oils for biodiesel production. Zeolite ANA and CAN are synthesized via hydrothermal processes by using diatomite as a raw material with an addition of proper alumina at 150 °C and 170°C for a certain duration. Various instruments such as XRD, SEM, FTIR and TPD are employed to characterize these zeolite catalysts. Process variables such as catalyst-to-oil, methanol-to-oil ratio, operation temperature are studied in order to optimize the reaction conditions.

Biodiesel is successfully produced via transesterification by using Zeolite ANA and CAN under appropriate conditions. As a consequence, a high yield above 90% from triolein / soybean oils to biodiesel can be achieved after a period of reaction time. Furthermore, both the spent Zeolite ANA and CAN could be regenerated with Na^{+}-impregnation and exhibited even better catalysis. However, in presence of free fatty acids, such as oleic acid in triglycerides, the yields decrease significantly. Therefore, a two-stage reaction strategy is adopted to process the acidic vegetable oils. (1) Zeolite ANA and CAN in H-form are firstly employed to esterify the oleic acid in feed oil and, subsequently, (2) transesterification of the remaining oils proceeds with the as-prepared zeolite ANA and CAN (in base form). A yield as high as ca. 90% is achieved with soybean oil having 2 wt% oleic acid (3.97 mg KOH / g oil) as the feed oil, zeolite ANA as the catalyst, and the reaction time with 3 h for esterification and 2h for transesterification.

Chemical looping gasification and hydrogen (CLGH) process from solid biomass in a dual fluidized-bed system

Jimin Zeng, Rui Xiao, Yihong Wang, Shanshan Shao and Dewang Zeng*

Key Laboratory of Energy Thermal Conversion and Control, Ministry of Education, School of Energy and Environment, Southeast University, Nanjing, 210096, P.R. China

E-mail: ruixiao@seu.edu.cn

Abstract

Chemical looping gasification (CLG) process is one of CO_2 capture technologies for producing syngas with high system conversion efficiency and inherent separation of CO_2. Chemical looping hydrogen (CLH) process with renewable energy sources is the other promising technology for producing H_2 and decreasing emission of greenhouse gas. Chemical looping gasification and hydrogen (CLGH) process from solid biomass is a promotion of the existing steam-iron process by combining CLG process and CLH process in one step.

This process in a dual fluidized-bed system can not only avoid highly concentrated CO_2 gas but also generate hydrogen with no direct efficiency loss. Meanwhile, the system offers two separate fluidized-bed reactors for biomass gasification and H_2 production. Iron-based oxygen carrier is used as bed material in the two fluidized beds to provide with lattice oxygen for gasification in the fuel reactor (FR) and to create a reduced environment for H_2 production in the steam reactor (SR). In this study, the experiment for CLGH process of pine sawdust was conducted in a dual fluidized-bed system, aiming to a suitable H_2/CO-ratio syngas and the best H_2 yield for the whole system. The effects of FR temperature, SR temperature, steam preheating temperature and steam-to-biomass ratio (S/B) on the process were investigated. The optimum operating conditions concluded in the experiment were: FR temperature at 820 °C, SR temperature at 910 °C, steam preheating temperature at 500 °C and S/B at 1.5 kg kg^{-1}. At this condition, the cold gas efficiency of 77.21 % and H_2 yield in the SR of 0.279 Nm^3 kg^{-1} were obtained. And the results enhance understanding of CLGH process and facilitate commercial utilization of solid biomass.

Key words

Chemical looping; Gasification; Hydrogen production; A dual fluidized-bed system; Biomass

1 Introduction

Biomass is a renewable energy source with the potential to be as a main substitute of fossil fuels. The main motivation to increase the use of this resource is for the longer availability and the security of the energy supply. Currently, biomass can be converted to high valuable energy forms via a number of processes including thermal, biological, and mechanical or physical processes. Since the increasing demand for gaseous fuels, gasification and hydrogen production have receive great attention and been extensively investigated by many researchers (Corella, Toledo et al. 2004, Li, Grace et al. 2004, Matsuoka, Kuramoto et al. 2008, Kern, Pfeifer et al. 2013, Woolcock and Brown 2013). Gasification is a chemical process that converts biomass into useful convenient gaseous fuels or chemical feedstock, the product gas from the gasifier is usually called syngas (mainly contains H_2, CO, H_2O, CH_4 and CO_2). Compared with combustion product gas, syngas has useful heating value, so gasification packs energy into chemical bonds while combustion releases it. Meanwhile, the purpose of gasification is not just energy conversion; production of chemical feedstock is also an important application. Research activities on hydrogen production increase rapidly since it is an environmental friendly fuel that does not emit carbon dioxide which is known as one of the global warming gases. In particular, hydrogen as a fuel for the mobile and stationary power generation by using fuel cell has a great potential to increase hydrogen demand in the near future (Cho, Kim et al. 2015).

Chemical looping is one of several emerging technology options capable of facilitating the uptake of low-emission energy technologies and helping in a diverse range of applications for productions of chemicals, fuels, or electricity. In the chemical looping process, a given reaction is divided into multiple sub-reactions, with each being typically carried out in a separate reactor (Moghtaderi 2011):

$$A + Me_xO_y \rightarrow C + Me_xO_{y-1} \quad (1)$$

$$B + Me_xO_{y-1} \rightarrow D + Me_xO_y \quad (2)$$

Overall:

$$A + B \rightarrow C + D \quad (3)$$

The link between sub-reactions is provided by oxygen carrier (OC), and the OC particles are reduced and regenerated in a cyclic fashion through the progress of the sub-reactions.

Chemical looping combustion (CLC) is one of the techniques used to combine fuel combustion and CO_2 capture. It is of growing interest in this technology due to the lower energy penalty and low cost equipment for CO_2 separation from flue gas (Xiao, Chen et al. 2012). Currently, the CLC process was investigated in TGA, a fixed bed or a fluidized-bed reactor with different OC particles (Shen, Wu et al. 2009, Song, Shen et al. 2013, Lyngfelt 2014). Xiao and Zhang et al. in our group did much work of this process (Xiao, Chen et al. 2012, Zhang, Xiao et al. 2013, Zhang, Xiao et al. 2013) , but combustion product gas did not have useful heating value. Considering the application of the product gas, advanced chemical-looping-based concept was proposed (Acharya, Dutta et al. 2009, Li, Kim et al. 2009, Moghtaderi 2011). Although the research of chemical looping hydrogen (CLH) process for H_2 was studied in our group (Xiao, Zhang et al. 2014), the process was complicated for commercial use.

In this work, chemical looping gasification and hydrogen (CLGH) process was proposed (Fig.1) and was conducted in a dual fluidized-bed system (Fig. 2). The reactor offered two separate fluidized bed reactors, fuel reactor (FR) and steam reactor (SR), for biomass gasification and hydrogen production. Iron ore was used as oxygen carrier (OC) for the circulation in the system.

In FR, the OC particles reacted with sawdust particles for gasification step:

$$(x+y)Fe_2O_3 + yC \rightarrow yCO + 2xFe + 2yFeO \quad (4)$$

$$(x+y)CO + (3x+y)Fe_2O_3 \rightarrow 2xFe_3O_4 + 2yFeO + (x+y)CO_2 \quad (5)$$

The reduced OC particles were transferred to AR, reacting with H_2O for generating H_2 (Cho, Kim et al. 2015):

$$3Fe + 4H_2O \rightarrow 4H_2 + Fe_3O_4 \quad (6)$$

$$3FeO + H_2O \rightarrow H_2 + Fe_3O_4 \quad (7)$$

Then, the OC particles were transported to the loop seal, where the oxidation reactions occurred:

$$4Fe_3O_4 + O_2 \rightarrow 6Fe_2O_3 \quad (8)$$

The OC particles between these two reactors transformation achieved the circulation of the system. The effect of FR temperature, SR temperature, S/B ratio and steam preheating temperature were extensively investigated in the work, in terms of gas distribution, H_2/CO ratio of syngas, carbon conversion, cold gas efficiency and H_2 yield.

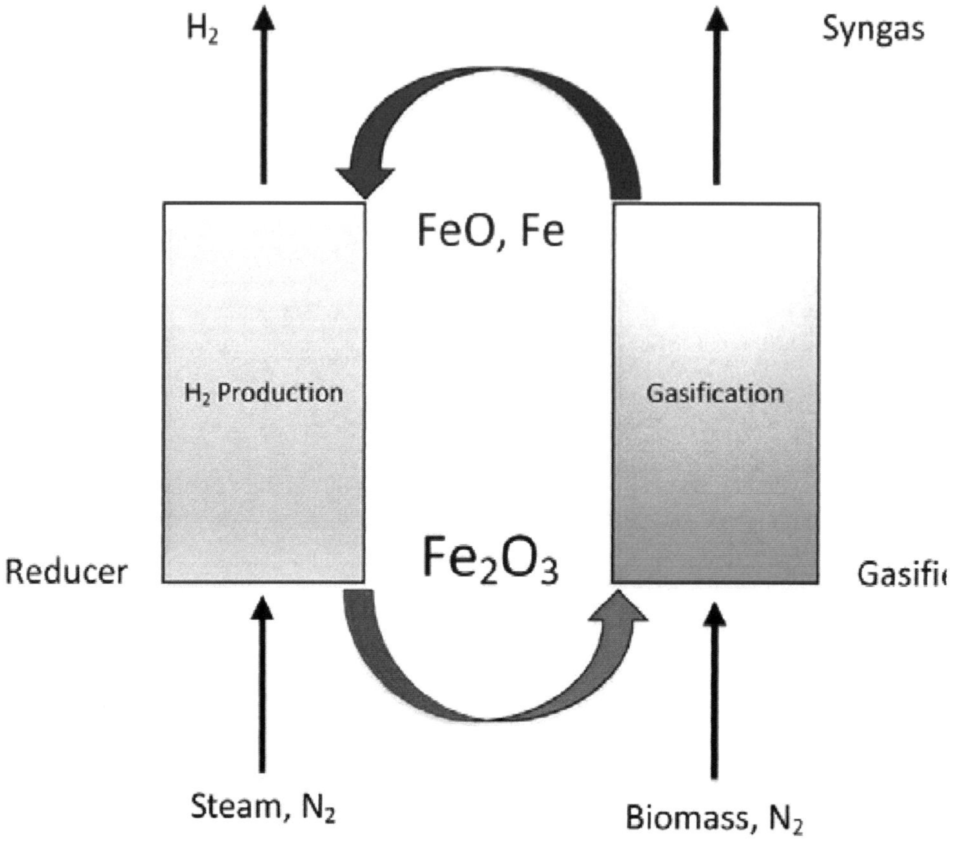

Fig.1. Schematic of chemical looping gasification and hydrogen (CLGH) process

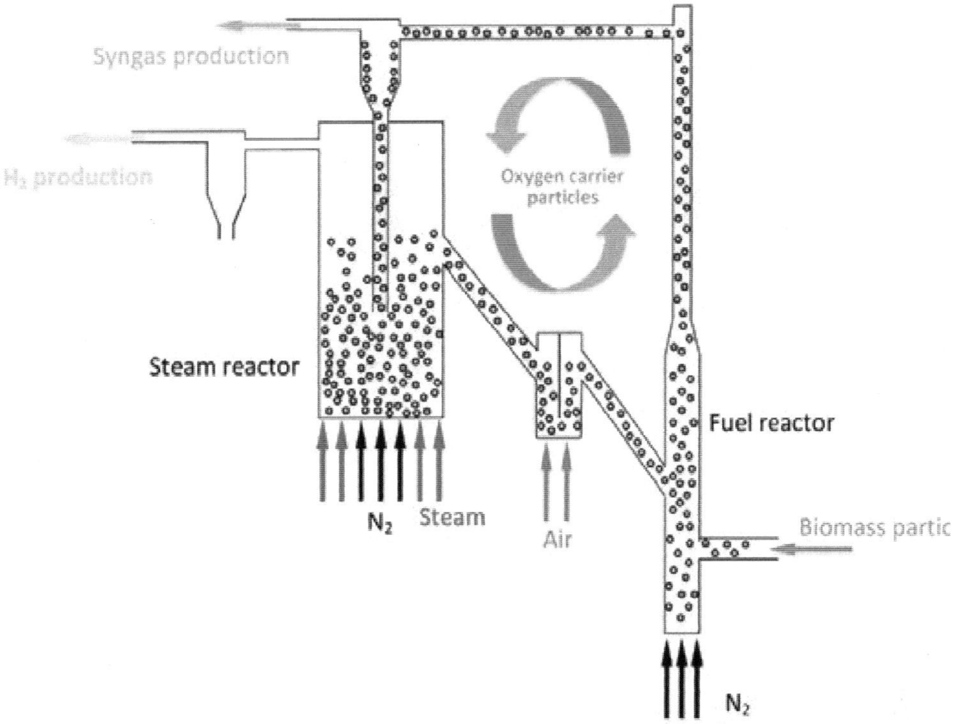

Fig.2. Schematic diagram of the dual fluidized-bed system for CLGH process

2 Materials and Methods

2.1. Material preparation and characterization.

Pine sawdust being obtained from Jiangsu Province (China) was used as feedstock in the tests. The sample was crushed and sieved into particles with a size range of 0.2 mm to 0.6 mm, and then was dried in a 105 °C oven for 12 h before experiment. Proximate and ultimate analyses were shown in Table 1.

Table 1. Proximate and ultimate analyses of pine sawdust

Parameter	Value
Proximate analysis (ad, wt.%)	
Moisture	1.44
Volatile	80.78
Fixed carbon	16.28
Ash	1.50
Ultimate analysis (ad, wt.%)	
C_{ad}	44.90
H_{ad}	6.38
$O_{ad}{}^{a}$	44.98
N_{ad}	3.74

Low heating value (MJ/kg)	18.33

ad: air dried basis.

[a] Calculated by difference

The OC particles used in this study was Companhia Valedo Rio Doce (CVRD) iron ore produced from Brazil, which was imported and provided by the NanjingMeishan Steel, Shanghai Baosteel Co. Ltd. The CVRD iron ore particles showed good performance compared to several other Australian iron ores during previous work in our group (Xiao, Song et al. 2010). The particles size range was within 0.1 mm to 0.3 mm.

2.2. Experimental Setup and Procedure

The experiment was conducted in a dual fluidized-bed system (Fig. 2) that included mainly a FR and a SR, a loop seal, hot gas cooler, biomass feeding unit, water/gas feeding unit, a steam generator, temperature control unit, and gas analysis system. The gas was analyzed online by MRU gas analyzers.

FR was a fast fluidized bed in diameter of 45 mm contracting to 30 mm in the height of 550 mm, and the total height was 1700 mm. SR was a bubbling fluidized bed with a circular column of 89 mm in diameter and 900 mm in height. The outlets of FR and SR were linked by cyclones. And FR and SR were connected by a rectangular loop seal with a cross section of 40×80 mm^2 and a height of 160 mm. Biomass feeding unit was 100 mm in height fixing at the right bottom of FR, and sawdust was fed into the reactor by two continuous feeding screw feeders. The first screw feeder was applied to definite quantitative determination. The secondary screw feeder was applied for feeding the feedstock in the reactor rapidly. The hopper was swept with N_2 to maintain an inert environment of the feeding unit. The reaction gases were controlled by Rota meters and preheated by heating tape to 250 °C before they were introduced into the reactors. The steam generator was an intelligent control water pump for generating steam through a preheating furnace. The entrance for the steam is fixed on the right of the fuel reactor at 600 mm in height.

Table 2 gave an overview of the main operating parameters of the experiments presented within this work. FR temperature and SR temperature were the temperature measured in the vertical middle of each fluidized bed.

Table 2. Overview of the operating parameters of experiments

	Unit	value
Iron ore (bed material)	kg	4
Feeding rate	kg h^{-1}	0.12
FR temperature	°C	700 to 880
SR temperature	°C	700 to 970

Steam-to-biomass (S/B) ratio	kg kg^{-1}	0.50 to 3.00
Steam preheating temperature	^{o}C	300 to 800

2.3. Data Evaluation

The outlet dry syngas consists mainly of H_2, CO, CO_2, CH_4, and N_2, since H_2O was removed. With the gas volume fractions measured by gas analyzer and the feeding N_2 balance gas flow rate, the outlet molar flow rate could be calculated based on nitrogen mass balance:

$$\dot{n}_{out} = \dot{n}_{in}\left(\frac{x_{N_2,in}}{x_{N_2,out}}\right) = \dot{n}_{in}\left(\frac{x_{N_2,in}}{1 - x_{H_2} - x_{CO} - x_{CO_2} - x_{CH_4}}\right) \quad (9)$$

where $\dot{n}_{in}$ was the total inlet gas molar flow rate; x_i was the volume fraction of gas species i measured by gas analyzer; $x_{i,in}$ and $x_{i,\text{out}}$ were the volume fractions of gas species i entering and exiting the reactor respectively.

The gas relative concentration (c_i) of each run in fuel reactor was calculated as:

$$c_i = \frac{x_i}{\sum x_i} \times 100\% \quad (10)$$

Carbon conversion efficiency (η_c) in the system was defined as the proportion of the carbon converted into gaseous products from the total carbon in the sawdust fed into FR:

$$\eta_c = \frac{12[(V_{CO} + V_{CO_2} + V_{CH_4}) \times \mathrm{Y}_{sg} + (V_{CO_2,hg} + V_{CO,hg} + V_{CH_4,hg}) \times Y_{hg}]}{22.4 \times M_C} \times 100\% \quad (11)$$

where Y_{sg} (Nm^3 kg^{-1}), Y_{hg} (Nm^3 kg^{-1}) and M_C (%) were the gas yield of the syngas production, the gas yield of hydrogen production and carbon fraction in the biomass, respectively. V_{CO} (%), V_{CO_2} (%) and V_{CH_4} (%) represented the volume fraction of CO, CO_2, CH_4 in the syngas, respectively, and $V_{CO_2,hg}$ (%), $V_{CO,hg}$ (%) and $V_{CH_4,hg}$ (%) represented the volume fraction of CO_2 ,CO, CH_4 in the hydrogen production gas.

The lower heating value (LHV, kJ Nm^{-3}) of the gas products is calculated as (Niu, Huang et al. 2013):

$$LHV = 126V_{CO} + 108V_{H_2} + 359V_{CH_4} + 635V_{C_2H_m} \quad (12)$$

where V_{CO}, V_{H_2}, V_{CH_4} and $V_{C_2H_m}$ were the volume fractions of CO, H_2, CH_4 and C_2H_m in the syngas, respectively.

Cold gas efficiency (η_e) was defined as the ratio of the heating value of the gas products from unit mass biomass gasification and hydrogen production to the total heating value of the unit mass of biomass:

$$\eta_e = \frac{LHV \times Y_{sg}}{q_{biomass}} \times 100\% \quad (13)$$

where Y_{sg} (Nm^3 kg^{-1}) was the gas yield of the production. LHV (kg Nm^{-3}) and $q_{biomass}$ (kJ kg^{-1}) were the lower heating value of the syngas and total heating value of biomass at room temperature, respectively.

Hydrogen yield (γ_{H_2}) was defined as the accumulated amount of hydrogen (Nm^3) produced in the steam oxidation period divided by the total amount of dry sawdust (kg) added into the reactor:

$$\gamma_{H_2} = \frac{22.4 \times x_{H_2,hg}}{Q_{biomass}} \quad (14)$$

where $x_{H_2,hg}$ (kmol h^{-1}) was the fraction of H_2 in SR and $Q_{biomass}$ (kg h^{-1}) represented the added biomass.

3 Results and Discussion

3.1. Effect of FR temperature.

FR temperature significantly affected the CLHG process in the dual fluidized-bed system. In the present work, the temperatures varied from 700 to 880 °C in 30 °C increments. AR temperature, S/B ratio and steam preheating temperature were kept constant at 910 °C, 1.50 kg kg^{-1} and 500 °C, respectively. The test results under different temperatures were presented in Fig. 3 and Table 3.

Fig. 3a showed the overall cumulative syngas composition on a dry basis at different FR temperatures. With the increase of the temperature, H_2 fraction increased from 42.47 % to 49.24 %. CO fraction slightly increased from 31.51 % to 33.13 %, while CO_2 fraction decreased from 13.70 % to 6.08 %. Table 3 showed cold gas efficiency of syngas and hydrogen yield of AR changed with different FR temperatures. Cold gas efficiency increased from 70.85 % to 77.21 % at 820 °C, and then decreased to 74.44 at a higher temperature. hydrogen yield was kept at the range of 0.256 $Nm^3\ kg^{-1}$ to 0.286 $Nm^3\ kg^{-1}$.

Results obtained from the experiments indicated that with the increase of FR temperature, biomass gasification reactions and became stronger, produced more H_2 and CO. Therefore, different amount of carbon was transferred to SR, which led to the increase of hydrogen yield in SR (from 0.256 to 0.286 $Nm^3\ kg^{-1}$). At a higher temperature (880 °C), Boudouard reaction in FR became stronger, which resulted in the decrease of hydrogen yield (Narvaez, Orio et al. 1996) .

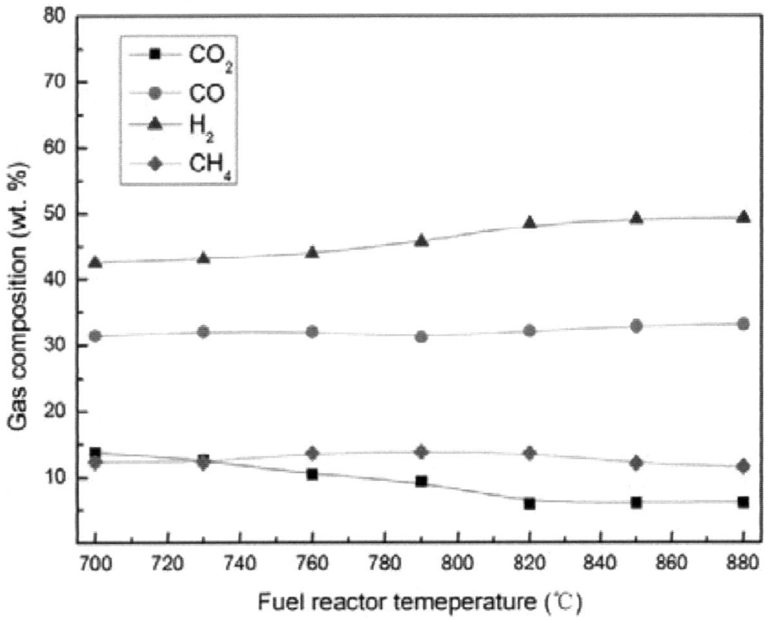

Fig.3. Effect of FR temperature on outlet gas composition of FR during reaction period

Table 3. Effect of FR temperature of the CLHG process

Temp. (°C)	Cold gas efficiency in FR (%)	Hydrogen yield ($Nm^3\ kg^{-1}$)
700	70.85	0.260
730	71.40	0.257
760	74.54	0.256
790	75.32	0.277

820	77.21	0.279
850	75.13	0.286
880	74.44	0.277

3.2. Effect of SR temperature.

Fig. 4 demonstrated the effect of SR temperature on cold gas efficiency of syngas and hydrogen yield in SR. The SR temperature increased from 700 °C to 970 °C, and FR temperature, S/B ratio and steam preheating temperature were set to 820 °C, 1.50 kg kg^{-1} and 500 °C, respectively. As indicated in Fig. 4, hydrogen yield increased with SR temperature on the stream, and reached the maximum of 0.279 Nm^3 kg^{-1} at 910 °C, and then stayed approximately constant at higher SR temperatures. Cold gas efficiency slightly decreased from 81.56 % to 74.88 %.

Considering the reactions in the SR (Reaction 6 and Reaction 7), when SR temperature increased, these reactions became stronger, which led to the increase of hydrogen yield. Cold gas efficiency slightly decreased for the stronger reactions of hydrogen production, and more amount of carbon were transported from FR to SR.

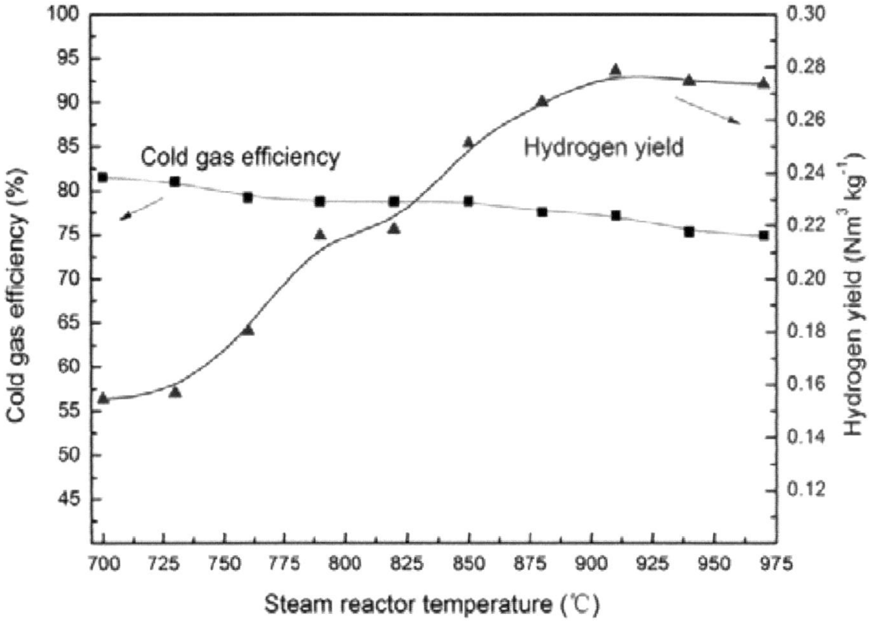

Fig.4. Effect of SR temperature on cold gas efficiency of syngas and hydrogen yield in SR

3.3. Effect of steam-to-biomass (S/B) ratio.

Fig. 5 showed the effect of S/B ratio in gas composition and cold gas efficiency of syngas and hydrogen yield in SR. As shown in Fig. 5, H_2 fraction increased from 37.15 % to 53.45 % while CO fraction decreased from 37.50 % to 33.03 % with the increase of S/B ratio. Cold gas efficiency remained constant first, and decreased at higher S/B ratios (> 1.50 kg kg^{-1}). Hydrogen yield increased rapidly from 0.50 to 1.75 kg kg^{-1}, and then slowed down at higher S/B ratios.

It should be noted that S/B ratio have a strong effect on hydrogen production process, the results demonstrated that the increase of the S/B ratio let the reactions in SR become stronger, while it would improve the amount of H_2O transported to FR through the circulation of the OC particles. It resulted in the water-gas reaction and shift reaction as shown below (Dennis, Scott et al. 2006, Matsuoka, Kuramoto et al. 2008):

$$C + H_2O \rightarrow CO + H_2 \quad (14)$$

$$C + 2H_2O \rightarrow CO_2 + 2H_2 \quad (15)$$

$$CO + H_2O \rightarrow CO_2 + H_2 \quad (16)$$

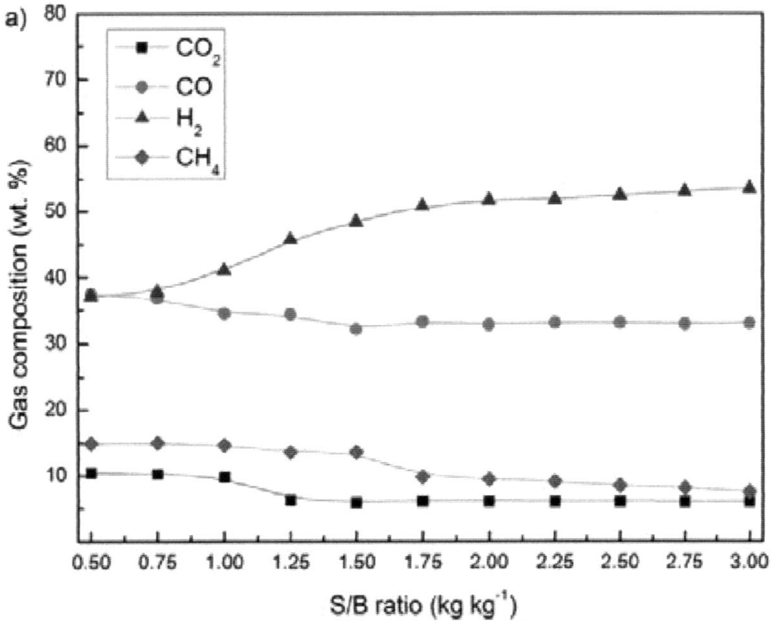

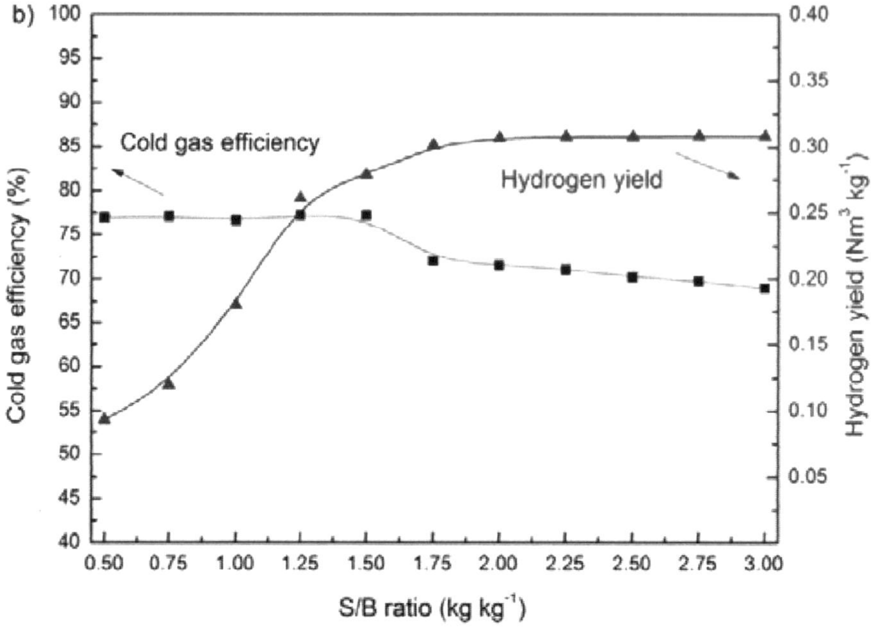

Fig.5. Effect of S/B ratio on a) gas composition of syngas and b) cold gas efficiency and hydrogen yield in SR during the reaction period

3.4. Effect of steam preheating temperature.

Table 4 illustrated the effect of steam preheating temperature of SR on cold gas efficiency of syngas and hydrogen yield. In Table 4, cold gas efficiency remained at approximately 77%, but hydrogen yield increased from 0.163 to 0.283 Nm^3 kg^{-1} with the increase of steam preheating temperature.

Since SR temperature was higher than the steam, the steam would absorb less heat from the reactor when increasing the steam preheating temperature, which resulted in the equilibrium of the reactions in SR, but it could slightly affected the gasification process in FR because of the circulation of the dual fluidized-bed system.

From the above investigations, the optimal conditions for the CLGH process of sawdust in the dual fluidized-bed system were: FR temperature of 820 ^{o}C, SR temperature of 910 ^{o}C, S/B ratio of 1.50 kg kg^{-1} and steam preheating temperature of 500 ^{o}C. At this condition, the cold gas efficiency of 77.21 % and H_2 yield in the SR of 0.279 Nm^3 kg^{-1} were obtained.

Table 4. Effect of steam preheating temperature of the CLGH process

Temp. (^{o}C)	Cold gas efficiency in FR (%)	Hydrogen yield (Nm^3 kg^{-1})

300	77.69	0.163
400	77.42	0.236
500	77.21	0.279
600	76.58	0.277
700	75.34	0.283
800	74.03	0.283

4 Conclusion

Chemical looping gasification and hydrogen (CLGH) tests using sawdust as fuel and iron ore as oxygen carrier were conducted in a dual fluidized-bed system. This process achieved gasification and hydrogen production at the same time. The following conclusions could be drawn from this study:

Significant effects occurred in the system. FR temperature influenced the gas composition of gasification process, while SR temperature and steam preheating temperature had strong impact on the hydrogen yield. S/B ratio affected gasification process and hydrogen production process at the same time because of the circulation in the whole system.

The optimum operation in the system was found as: FR temperature of 820 ^{o}C, SR temperature of 910 ^{o}C, S/B ratio of 1.50 kg kg^{-1} and steam preheating temperature of 500 ^{o}C. At this condition, 77.21 % of cold gas efficiency in syngas and 0.279 Nm^3 kg^{-1} of hydrogen yield obtained for the process. Based on this work, the CLGH process will allow for determining the most promising options for gasification and hydrogen production processes.

Acknowledgement

The authors gratefully acknowledge the financial support by National Key Basic Research Program of China (973 Program), (Grant No. 2012CB215306) and National Natural Science Foundation of China (NSFC), (Grant No. 51476035).

Reference

Acharya, B., et al. (2009). "Chemical-looping gasification of biomass for hydrogen-enriched gas production with in-process carbon dioxide capture." Energy & Fuels **23**(10): 5077-5083.

Cho, W. C., et al. (2015). "Activation and reactivity of iron oxides as oxygen carriers for hydrogen production by chemical looping." Industrial & Engineering Chemistry Research **54**(12): 3091-3100.

Corella, J., et al. (2004). "Olivine or dolomite as in-bed additive in biomass gasification with air in a fluidized bed: which is better?" Energy & Fuels **18**(3): 713-720.

Dennis, J., et al. (2006). "In situ gasification of coal using steam with chemical looping: a technique for isolating CO_2 from burning a solid fuel." Journal of the Energy Institute **79**(3): 187-190.

Kern, S., et al. (2013). "Gasification of wood in a dual fluidized bed gasifier: Influence of fuel feeding on process performance." Chemical Engineering Science **90**: 284-298.

Li, F., et al. (2009). "Syngas chemical looping gasification process: oxygen carrier particle selection and performance." Energy & Fuels **23**(8): 4182-4189.

Li, X., et al. (2004). "Biomass gasification in a circulating fluidized bed." Biomass and bioenergy **26**(2): 171-193.

Lyngfelt, A. (2014). "Chemical-looping combustion of solid fuels–status of development." Applied Energy **113**: 1869-1873.

Matsuoka, K., et al. (2008). "Steam gasification of woody biomass in a circulating dual bubbling fluidized bed system." Energy & Fuels **22**(3): 1980-1985.

Moghtaderi, B. (2011). "Review of the recent chemical looping process developments for novel energy and fuel applications." Energy & Fuels **26**(1): 15-40.

Narvaez, I., et al. (1996). "Biomass gasification with air in an atmospheric bubbling fluidized bed. Effect of six operational variables on the quality of the produced raw gas." Industrial & Engineering Chemistry Research **35**(7): 2110-2120.

Niu, M., et al. (2013). "Simulation of syngas production from municipal solid waste gasification in a bubbling fluidized bed using Aspen Plus." Industrial & Engineering Chemistry Research **52**(42): 14768-14775.

Shen, L., et al. (2009). "Chemical-looping combustion of biomass in a 10 kWth reactor with iron oxide as an oxygen carrier." Energy & Fuels **23**(5): 2498-2505.

Song, T., et al. (2013). "Evaluation of hematite oxygen carrier in chemical-looping combustion of coal." Fuel **104**: 244-252.

Woolcock, P. J. and R. C. Brown (2013). "A review of cleaning technologies for biomass-derived syngas." Biomass and bioenergy **52**: 54-84.

Xiao, R., et al. (2012). "Pressurized chemical-looping combustion of coal using an iron ore as oxygen carrier in a pilot-scale unit." International Journal of Greenhouse Gas Control **10**: 363-373.

Xiao, R., et al. (2010). "Pressurized chemical-looping combustion of coal with an iron ore-based oxygen carrier." Combustion and Flame **157**(6): 1140-1153.

Xiao, R., et al. (2014). "Use of heavy fraction of bio-oil as fuel for hydrogen production in iron-based chemical looping process." International Journal of Hydrogen Energy **39**(35): 19955-19969.

Zhang, S., et al. (2013). "Performance of Fe_2O_3/$CaSO_4$ composite oxygen carrier on inhibition of sulfur release in calcium-based chemical looping combustion." International Journal of Greenhouse Gas Control **17**: 1-12.

Zhang, S., et al. (2013). "CO_2 Capture and Desulfurization in Chemical Looping Combustion of Coal with a $CaSO_4$ Oxygen Carrier." Chemical Engineering & Technology **36**(9): 1469-1478.

404872 Modular Production of Synthetic Petroleum from Shale Gas

Abstract:

A current issue related to the production of petroleum from shale formations is the co-production of natural gas. In many cases, this co-produced gas is flared since the infrastructure is not in place to allow transport. The flaring operations produce carbon emissions and make no use of the energy content of the natural gas. The possibility exists to convert this flared natural gas to a valuable product using Fischer Tropsch fuel synthesis technology.

Ceramatec is currently operating a small, ¼ BPD Gas-to-Liquids pre-pilot plant at its Salt Lake City, Utah facility. The existing laboratory system has been run with a variety of catalysts and a variety of conditions. Based on the test results, a modular GTL plant that is road transportable has been designed and fabrication cost estimated on a per barrel-per-day (BPD) of capacity. The paper will describe the test results that have been obtained to date and a series of planned tests in association with a ~2 BPD CTL program that is funded by US Department of Energy. Further, the paper will discuss the cost structure associated with modular GTL plants of various sizes and determining factors for those costs.

420207 Optimal Production of Light Olefins from "Wet" Shale Gas: An Integrated NGL Cracking and Dry Gas Reforming Approach

Abstract:

High levels of natural gas production that stemmed from the shale gas industry lowered this feedstock price significantly [1]. Therefore, production of "wet" gas that contains natural gas liquids (NGLs), such as ethane, propane, butane, and natural gasoline, is favorable to extract for commercial purposes [1]. In some wet shale plays, the combined ethane and propane composition can exceed 30% [2]. These recent trends motivate industrial effort to build several new ethane crackers with a combined ethylene production capacity of 12.5 million tonnes/year in the United States [3].

Light olefins including ethylene, propylene, and butene isomers are valuable petrochemical intermediates and make up two-thirds of the petrochemicals market [4]. These olefins can be produced from alternative feedstocks such as biomass and natural gas via thermochemical conversion and reforming [4]. However, extraction of NGLs prior to methane conversion is an opportunity for refineries to maximize carbon conversion and avoid high reforming costs. A demethanizer column is utilized [5,6] to recover more than 83% of ethane and more than 99% of higher hydrocarbons in the NGL stream. The NGLs can be cracked or further separated for C_3 and C_4 dehydrogenation processes. A rigorous mathematical model is developed for the hydrocarbon cracker using the well-established cracking kinetics in the literature [7,8]. Dry gas is converted via reforming or direct conversion processes introduced into a process superstructure [4,9]. The light olefins are produced and purified to product quality with an extensive superstructure that considers multiple production, conversion, and purification technologies.

The process superstructure that is developed is solved using a novel branch and bound global optimization framework. The objective is to maximize the profit of light olefin production from the integrated NGL extraction and dry gas conversion plant. A techno-economic analysis will be presented with major topological decisions across several case studies. Different natural gas compositions will be shown to present the topological trade-offs for various NGL compositions.

References:

[1]: Kopalek, M.; High value of liquids drives U.S. producers to target wet natural gas resources, 2014, Energy Information Administration

[2]: Hill, R. J.; Jarvie, D. M.; Zumberge, J.; Henry, M.; Pollastro, R. M.; Oil and gas geochemistry and petroleum systems of the Fort Worth Basin, 2007, AAPG Bulletin, 91(4), 445-473.

[3]: Chang, J.; New projects may raise US ethylene capacity by 52%, PE by 47%, 2014, ICIS Petrochemicals.

[4]: Onel, O.; Niziolek, A. M.; Elia, J. A.; Baliban, R. C.; Floudas, C. A.; Biomass and Natural Gas to Liquid Transportation Fuels and Olefins (BGTL+C2_C4): Process Synthesis and Global Optimization, 2015, Industrial and Engineering Chemistry Research, 54, 359-385.

[5]: Luyben, W. L.; NGL Demethanizer Control, 2013, Industrial and Engineering Chemistry Research, 52, 11626-11638.

[6]: Luyben, W. L.; Effect of Natural Gas Composition on the Design of Natural Gas Liquid Demethanizers, 2013, Industrial and Engineering Chemistry Research, 52, 6513-6516.

[7]: Sundaram, K. M.; Froment, G. F.; Modeling of thermal cracking kinetics – I: Thermal cracking of ethane, propane and their mixtures, 1977, Chemical Engineering Science, 32(6), 601-608.

[8]: Sundaram, K. M.; Froment, G. F.; Modeling of thermal cracking kinetics – II: Cracking of iso-butane, of n-butane, and of mixtures ethane-propane-n-butane, 1977, Chemical Engineering Science, 32(6), 609-617.

[9]: Baliban, R. C.; Elia, J. A.; Floudas, C. A.; Novel Natural Gas to Liquids Processes: Process Synthesis and Global Optimization Strategies, 2013, AIChE Journal, 59(2), 505-531.

OPTIMIZING LNG PRODUCTION VIA BOIL-OFF GAS SYSTEM DESIGN

John Ergina, Technical Professional - Process
Sumit Sharma, Senior Technical Professional - Process
KBR Inc., Houston, Texas, USA

1. Introduction

The economic storage of liquefied natural gas (LNG) requires near-atmospheric conditions in LNG product storage tanks. In a typical LNG process, the natural gas stream is liquefied at high pressure in a cryogenic heat exchanger. The pressure of the LNG product stream from the cryogenic heat exchanger is then reduced to enable low pressure storage in the LNG tanks. Low pressure storage operation and equipment generate flash vapor collectively termed as Boil-Off Gas (BOG). BOG generation is a continuous process in LNG liquefaction plants. Local regulations, environmental concerns, and economic considerations prevent flaring as a means to dispose of BOG. Instead, BOG is either used as fuel gas or recycled to the inlet of the LNG train for liquefaction.

Historically, LNG facilities have been developed in challenging and remote locations with limited or no access to modern infrastructure. These facilities are designed to be self-sufficient, especially for power generation. BOG is an important fuel gas source in LNG plants for the gas turbine drivers used for refrigerant compressors and electrical power generation. A self-sufficient LNG plant requires well-integrated systems where the amount of BOG is controlled to satisfy a significant portion of the fuel gas demand for the facility. The current opportunity for LNG exports in North America, a continent with highly developed electrical infrastructure, has provided new options for LNG plant design. Plants in North America have electric power available from municipal utility companies which can reduce the interconnectedness or complexity of the certain units of the LNG plant.

The existence of BOG invariably adds capital cost (CAPEX) to the LNG plant economics. BOG generation and its handling can also have a significant impact on LNG design and production rate. A robust design of the BOG system manages the added cost to provide additional value by optimizing net LNG production.

2. Boil-Off Gas: Sources and Users

BOG is primarily a result of storing LNG in tanks at near atmospheric pressure. Other sources of BOG generation include vapor displacement in tanks from incoming LNG product and heat leak into LNG tanks and piping due to ambient conditions or equipment operation. Estimating the BOG rate depends on the mode of LNG plant operation. An LNG plant can be classified as operating in Holding mode or Loading mode. During holding mode, or steady state operation with no ship loading, BOG is estimated from:

- LNG Tank (only) flash on LNG rundown
- LNG storage tank(s) heat leak (max allowed 0.05 volume% of a full tank's contents vaporized per day for onshore tanks, an assumption typically based on 100% methane)
- Vapor displacement from the storage tanks
- Heat leak in LNG rundown and loading lines

In ship loading operation mode, the following additional BOG sources are also included:

- Ship flash on loaded LNG due to LNG loading pump heat input (and ship cool down if arrives to the jetty warm)
- LNG ship heat leak (0.15 volume% of a full ship's contents vaporized per day)

The BOG generated in the storage tanks is compressed by the BOG compressors up to a sufficient pressure to route the BOG to the fuel gas system or the front of the liquefaction section. Part of the BOG vapor generated during the ship loading operation is returned to the storage tanks to replace the liquid volume removed by the pumps, with the excess gas going to the BOG compressors.

The BOG from the storage is utilized as fuel gas. The most significant users for the BOG are typically the gas turbines for the refrigeration compressor drivers and power generation.

3. Process Design Parameters

For a given LNG production rate, BOG system design and capacity can be affected by both feed gas composition and the overall plant design requirements. In response to the feed gas composition variation or plant fuel gas requirements, BOG production rate can be varied by adjusting the LNG product temperature exiting the cryogenic heat exchanger. If all other LNG product specifications are met, optimizing BOG system design also results in optimized refrigeration compressor load and LNG production rate.

In the following sections, three case studies are considered to illustrate the effect of feed gas composition and fuel gas demand on overall BOG system design. These case studies are based on a 5 Mt/a LNG production train using Air Products and Chemicals, Inc. (APCI) Propane Pre-Cooled Mixed Refrigerant (C3MR) ® process. For the case studies, dedicated end-flash gas collecting units in the LNG train, which assist in additional waste heat recovery, are not considered. For the purpose of the case studies, the BOG will be either:

- Recycled and mixed with the natural gas upstream of the liquefaction unit to avoid temperature fluctuations to the main heat exchanger
- Used as regeneration gas for the dehydration molecular sieve beds (before being used as fuel gas)
- Used as fuel gas directly, or
- A combination of above options

Optimization of LNG production will be evaluated with Specific Power (η), which is total compression power divided by LNG steady state mass flow:

$$\eta = \frac{P_{C3} + P_{MR} + P_{BOG} + P_{Feed}}{\dot{m}_{LNG}}$$

Where,

- P_{C3} = Propane Compressor Power
- P_{MR} = Mixed Refrigerant Compressor Power
- P_{BOG} = Boil-off Gas Compressor Power
- P_{Feed} = Feed Gas Booster Compressor Power

- $\dot{m}_{LNG}$ = Steady State Mass flow of LNG (in-Tank)

Effect of design choices based on feed gas composition and LNG plant fuel gas demand is examined and evaluated in terms of the change in LNG production and project economics.

3.1 *Case Study 1 - "Weight Watchers": Impact of Feed Gas Composition*

Feed gas to the LNG production facility can have a range of compositions, from lean feed gas (high concentration of low molecular weight / boiling point components) to rich feed gas (high concentration of high molecular weight / boiling point components). For LNG plant design, the feed gas envelope of a facility is critical for optimizing the BOG system. For a given feed gas rate and liquefaction temperature and pressure, any increase in concentration of low boiling point components will result in higher BOG generation from the flash in the LNG storage tanks. To handle the variation in feed gas composition, BOG system design should be flexible enough such that there is minimum impact on the LNG production rate.

In this case study, three feed gas compositions are considered. A Lean feed gas case (at 97 mol% methane concentration, Molecular weight: 16.48) is compared against an Average feed gas case (93 mol% methane concentration, Molecular weight: 17.18) and a Rich feed gas case (at 90 mol% methane concentration, Molecular weight: 17.69).

At constant LNG production and liquefaction operating conditions (temperature and pressure), the Lean feed gas generates the largest BOG flow among the three feed gas cases as shown in Table 3-1.

When LNG is flashed in the storage tank, there is a change in enthalpy because of the pressure change. Leaner gas with more methane and nitrogen (lower boiling point components) requires more cooling duty because of their higher tendency to be vapor at a constant liquefaction temperature and pressure than richer gas with ethane, propane, and butane.

Table 3-1: Feed gas composition effect on BOG

Case	BOG Volumetric Flow	BOG Mass Flow
Lean feed gas (97% methane)	100 %	100 %
Average feed gas (93% methane)	85 %	98 %
Rich feed gas (90% methane)	80 %	96 %

For the high BOG generated in the Lean feed gas case, the total power required to compress the BOG at the same compression ratio is also highest compared to that for the Average and Rich feed gas cases (Figure 3-1). In comparison, the Rich gas case generates about 20% less BOG and demands over 10% less compression power compared to the Lean gas case.

If the plant is designed based on the Lean feed gas case, the BOG compressor rated flow will be approximately 15% greater in capacity than the Average feed gas case along with additional electrical power required.

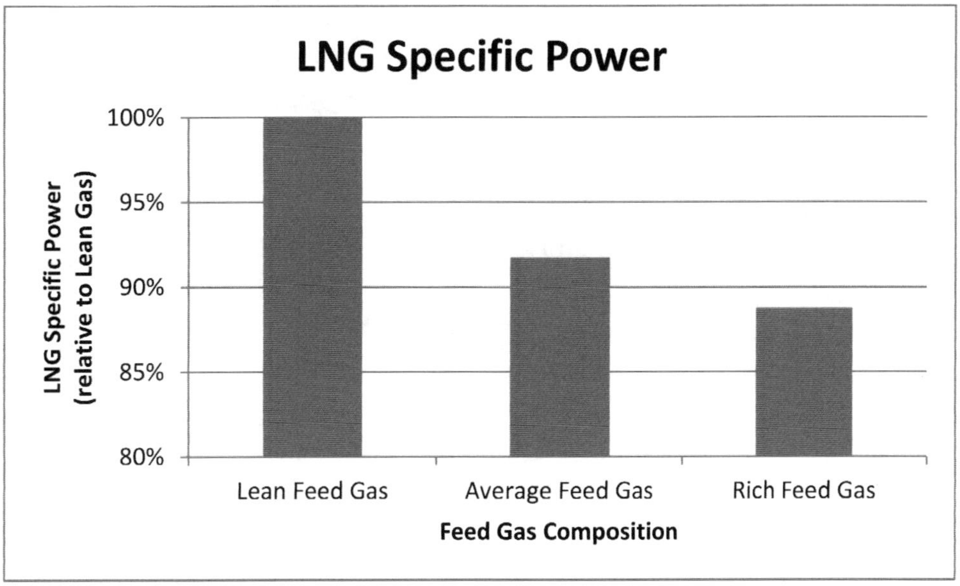

Figure 3-1: Difference in Plant LNG Specific Power

Typically, BOG Compressors with Inlet Guide Vane (IGV) control have a range of capacity control with good turndown efficiencies; therefore, the selection of the compressor guarantee point is important. A lean feed gas requires a large compressor and driver which will increase the plant CAPEX over a design based on a rich feed gas. This design will require 20% turndown during the Rich feed gas case, resulting in a reduction in compressor efficiency.

In contrast, if the Rich feed gas case is used for sizing the BOG system it would result in a small compressor and driver as well as low CAPEX. However, the low CAPEX approach is not designed to handle the additional BOG generated during the Lean feed gas case (at the same LNG production rate). The LNG plant will be required to turn down to about 96% of total target production (mass flow basis) to accommodate additional BOG generated during the Lean feed gas case, resulting in ~ 4.0% loss of the plant LNG production and significant financial impact for the LNG plant.

3.2 Case Study 2 - "What's Cooler than Ice Cold...LNG Cold": Impact of Cryogenic Heat Exchanger Outlet Temperature

Overall plant fuel gas demand is also an important consideration for the BOG system design. Consumption of the BOG as a fuel for various plant design requirements provides an estimate of the required BOG capacity. Self-sufficient LNG facilities with on-site power generation will have a large BOG unit to support a high fuel gas demand. However, LNG facilities with convenient access to a local and reliable electrical power grid for plant power, such as in the US Gulf Coast, will have low fuel gas demand and a small BOG unit. These two fuel gas consumption scenarios create an envelope for fuel gas consumption and BOG system design.

Total fuel gas demand in an LNG plant can be met primarily with BOG, along with treated feed gas as a complementary source. For a given LNG train capacity, the amount of BOG to be generated is dictated by the LNG product outlet temperature at the cryogenic heat exchanger. A (colder liquefaction temperature at the outlet of cryogenic heat exchanger reduces the BOG generation rate but increases the load on cryogenic heat exchanger and/or the refrigeration compressors. In contrast, a warmer temperature produces more BOG, with excess BOG recycled to inlet of the liquefaction unit. Re-liquefaction of this lower boiling point gas (nitrogen and methane mostly) increases the cryogenic heat exchanger load at a fixed outlet temperature. The Table 3-2 shows the difference in the associated power required at different cryogenic heat exchanger outlet temperatures. To illustrate the differences, the power requirements in Table 3-2 are based on Average feed gas composition (93% methane) and shown relative to the case with a cryogenic heat exchanger outlet temperature of minus 140°C.

Table Error! No text of specified style in document.**-2: Cryogenic Heat Exchanger Outlet Temperature and Compressor Power Distribution**

Case	BOG Compressor Power	Feed Gas Booster Compressor Power	Refrigerant Compressor Power	Overall Power
-140 °C	100 %	100 %	100 %	100 %
-145 °C	73 %	95 %	103 %	99.34 %
-150 °C	49 %	91 %	104 %	97.12 %
-155 °C	31 %	88 %	107 %	97.38 %
-160 °C	15 %	85 %	111 %	98.34 %

The capacity constraint in typical LNG plant designs is the required refrigeration capacity (cryogenic heat exchanger and/or the refrigerant compressors). Therefore, in general it is beneficial to reduce the refrigeration load. Table 3-2 also shows the relationship between the refrigerant compressors and BOG and Feed Gas Booster compressor power. As the cryogenic heat exchanger outlet temperature gets colder, the BOG and Feed Gas Booster compressor power decreases since the LNG is more sub-cooled and generates less, but the refrigerant compressor power increases because more cooling duty is required in the cryogenic heat exchanger. The trend of LNG Specific Power is shown in Figure 3-2.

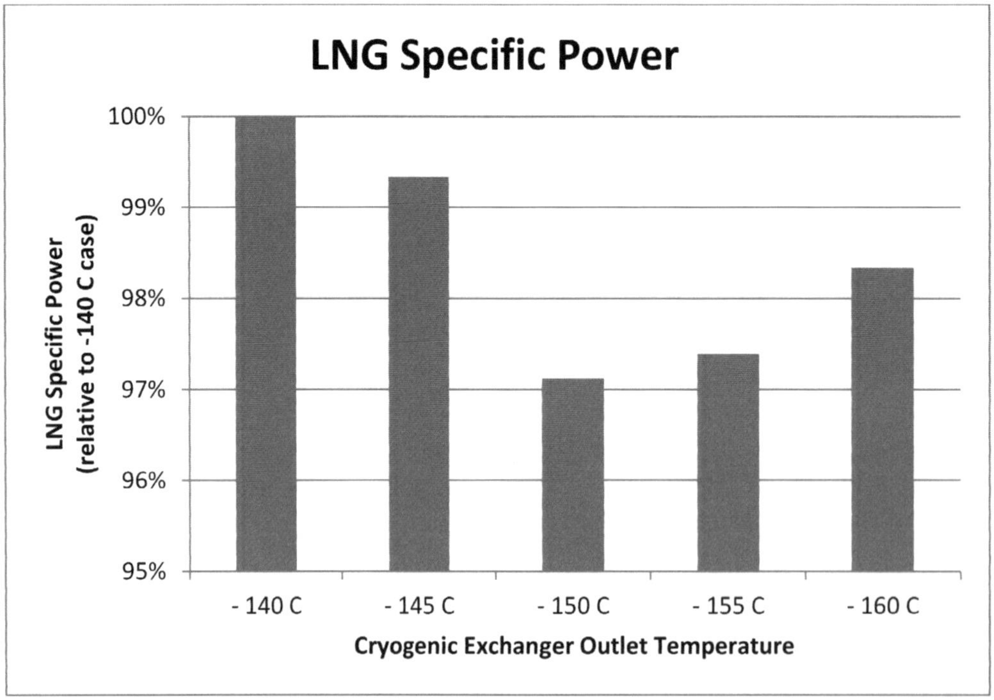

Figure 3-2: LNG Product Temperature and Plant LNG Specific Power

In this case study, as the cryogenic heat exchanger outlet temperature drops, the LNG conversion efficiency improves until it reaches a temperature around minus 150°C; if the temperature is pushed warmer or colder the specific power increases. For the warm outlet temperature cases, more and more BOG is generated that has to be recycled for re-liquefaction; this scenario also results in higher nitrogen content in the feed gas to the liquefaction section. Around a temperature of minus 150 °C, the plant performs at the optimum balance between BOG compression and refrigerant compression demand. Designing for the lowest total compression power optimizes the LNG production.

In the US Gulf Coast, the availability of municipal electric power provides the option of designing an all electric LNG (e-LNG) facility which provides benefit in plant fuel gas emissions In e-LNG, the Gas Turbine Generators (GTGs) and associated packaged equipment are eliminated and the Gas Turbine Drivers (GTDs) for the refrigeration compressors are replaced with electric motors. The fuel gas demand in the plant is also significantly reduced. The electrical grid removes the need for the GTGs and associated equipment/piping, resulting in reduced CAPEX; however, in order to accommodate the power import from local grid, new electrical equipment will be required for converting the grid based power to usable power for the LNG plant.

For an e-LNG design, the generated BOG can be managed by either reducing the cryogenic heat exchanger outlet temperature or increasing the BOG recycle rate. This is accomplished by shifting the outlet temperature to a colder temperature. Both of these options result in an increase of the LNG specific power (compared to a GT LNG plant) because of greater compressor power required to achieve the target LNG production. Furthermore, electric motor designs require adequate redundancy and sparing which can negatively impact the project economics even though the electric motor driver design provides more flexibility than the capacity specific gas turbine drivers.

The overall BOG system design and LNG production can be optimized for a either a gas turbine or e-LNG plant design but the results of the project economic analysis can sway project decisions.

4. Conclusion

The operational envelope of the feed gas composition and the cryogenic heat exchanger outlet temperature are important factors that affect the BOG system design. Optimization of the BOG system design and LNG production will require consideration of Lean feed gas composition and a cryogenic heat exchanger outlet temperature that minimizes LNG specific power. Designing the BOG system for a Lean feed gas case will avoid the economic impact of reducing LNG production to accommodate higher BOG from a leaner feed gas. Similarly, optimizing the distribution of compressor power between the BOG and refrigeration compressors by designing for the optimum cryogenic heat exchanger outlet temperature provides value by minimizing LNG Specific Power.

439894 In-situ multi-scale characterization of the adsorption and desorption behaviors of CO_2 with shale for sustainable extraction of unconventional hydrocarbons

Abstract:

Increasing demand for energy in the global economy has a detrimental impact on the overall sustainability of the planet. In the United States, shale gas and oil have emerged as significant energy resources. Current approaches to use water for hydrofracking are environmentally unsustainable in the long run and alternative fracturing fluids such as CO_2 need to be explored for sustainable energy extraction. The key factors to evaluate are the changes in the pore volume and the chemical compositions of shales when reacted with CO_2. Particularly, the in-situ adsorption and desorption behavior of CO_2 on shales and the corresponding chemical and structural changes in shales are important to investigate for the sustainable extraction of unconventional hydrocarbons. In this study, simultaneous in-situ measurements of the changes in the molecular structure, microstructure, and pore volume of shales during adsorption and desorption of CO_2 are performed using Wide, Small, and Ultra-Small Angle X-Ray Scattering (WAXS/SAXS/USAXS) to determine whether the chemical and structural changes are reversible or irreversible. In addition, the effects of CO_2 on the chemical and morphological changes of shales in the presence and absence of variable amounts of water at elevated temperatures and pressures (T_{max} = 80°C, P_{max} = 150 atm) showed that increasing the water content increases the pore spaces in shales by inducing chemical changes.

432874 Simulation and Optimization of CNG Mixtures Fill-up for Vehicles

Abstract:

First, a novel mathematical model is proposed, based on thermodynamics and transport phenomena fundamentals, that aims to capture the CNG mixture's pressure, temperature and molar volume evolution during a CNG vehicle's fill-up process. The CNG mixture's thermodynamic properties are calculated through the use of the generic cubic equation of state and residual properties. The obtained model gives rise to a set of differential-algebraic equations (DAE), which are then simulated using a hybrid Newton/Runge-Kutta method. The cooling behavior during the beginning of the fill-up is studied, as well as the heating behavior once the heating by compression mechanism takes place.

Afterwards, the fill-up process is formulated as a minimum time optimal control problem. The process of filling-up high pressure gas storage vessels consists of a gas source tank, an isenthalpic (Joule-Thomson or J-T) valve, a cooling system, and a gas storage vessel. These units are assumed to be thermally insulated. Despite the nonlinear nature of the aforementioned optimal control problem, its global solution is obtained analytically. A novel transformation technique is employed, to decompose the problem into a process simulation problem independent of time, and a simpler minimum time control problem that only depends on the final molar density value and the maximum allowable feed mass flowrate. The feasibility of the fill-up is uniquely determined by the process simulation problem, and upon fill-up feasibility, the minimum time control problem is then globally solved. Two fill-up case studies, involving two different system configurations are analyzed. In Case 1, the fill-up process has a constant molar enthalpy feed, and no cooling system. Case 2 considers a fill-up process with a constant temperature feed, delivered by an efficient cooling system. The obtained results for both case studies are compared to pure methane compression results.

Anticipating Leaner Feed Gas in Badak LNG Plant

Norvan Hery Kusnandar
Process Engineer
PT. Badak Natural Gas Liquefaction
Bontang, East Kalimantan, Indonesia
Phone: +62-548-553841
Fax: +62-548-552234
E-mail: norvan@badaklng.co.id

Background

Badak LNG plant considers receiving new feed gas sources from deepwater drilling and other various drilling projects in the future. It is anticipated most of these new feed gas sources will be significantly leaner than the current sources. As the new feed gas sources become available, the feed gas to the Badak LNG plant will vary in composition over time. These leaner feed gases may impact the Badak LNG plant operation with respect to LNG product quality, turndown capability of existing distillation columns, refrigerant make up requirement and may even have the potential freeze-out of benzene or heavy hydrocarbons with the current plant configuration.

Two scenarios (commingle and segregation) were previously developed to assess the optimum operation as well as commercial aspect both for upstream and downstream facilities. In order to have win-win solution to combine the advantages of both scenarios, a new scenario is proposed and called as 'commingle hybrid' mode. The commingle hybrid mode is presumed to give better result in term of LNG contract, additional LPG production, additional production revenue and less plant modification cost. Badak LNG will continue the plant operation in commingle hybrid mode for anticipating leaner feed gas.

Feed Gas Profile

The existing Badak LNG plant consists of eight natural gas liquefaction trains using the AP-C3MR LNG Process. The plant is fed with gas produced in different fields in East Kalimantan and routed to the plant through four pipelines. Starting 2016, some new gas fields are projected to deliver gas to Badak LNG plant. This new incoming gas is considered leaner than existing.

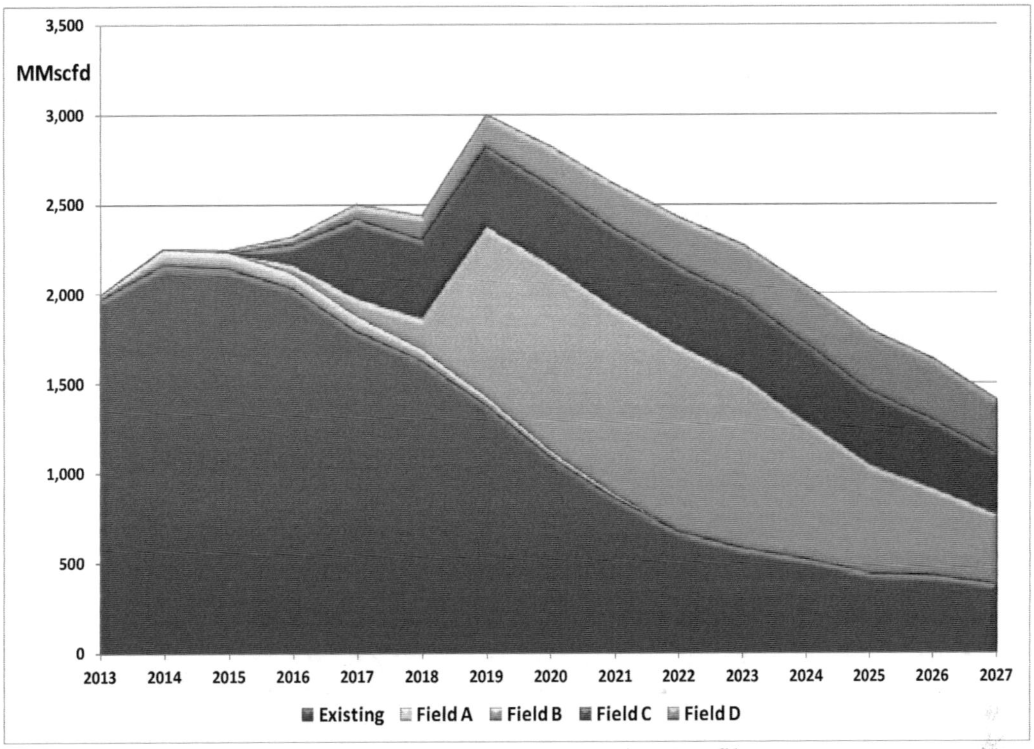

Figure 1. Feed Gas Production Profile

The new feed gas are mostly leaner than the existing which contains more methane and less CO_2 and BTEX (benzene, toluene, ethyl-benzene, xylene).

Table 1. Feed Gas Characteristic

Composition	Existing	Field A	Field B	Field C	Field D
N_2 (%)	0.10	0.23	0.15	0.04	-
CO_2 (%)	5.49	0.75	0.72	1.05	6.93
C_1 (%)	85.32	97.61	90.89	97.18	79.84
C_2 (%)	4.04	0.33	3.84	0.77	7.99
C_3 (%)	2.93	0.43	2.28	0.47	3.11
C_4 (%)	1.31	0.39	0.88	0.27	1.26
C_5+ (%)	0.83	0.36	0.34	0.22	0.34
Benzene (ppm)	232	-	33	13	131
Toluene (ppm)	179	-	15	13	112
Xylene (ppm)	25	-	2	22	11
Cyclohexane (ppm)	265	100	-	107	125

Plant Operation

It is hypothesized that there are 4 main parameters to be concerned dealing with leaner feed gas characteristic in regards to the impact to LNG plant operation. These parameters are LNG product quality, turndown capability of existing distillation columns, refrigerant make up requirement, and may even the potential freeze-out of benzene or heavy hydrocarbons with the current plant configuration. Therefore, four parameters are set based on its design data to assess the impact to the normal operation. Those parameters setting are :

a. LNG quality : minimum density of 453 kg/m^3 or equal with LNG heating value of 1,107 Btu/scf

b. Potential HC freezing : minimum C_5+/benzene/cyclohexane at 0.1 %/1.46 ppm/1.00 ppm, respectively, at gas inlet of LNG Main Heat Exchanger.
c. Availability of refrigerant make up : minimum C_2/C_3 make up requirement are 210/381 Nm^3/hour/train, respectively.
d. Fractionation turn down ratio : minimum at 40% for all fractionation columns (Scrub Column/Deethanizer/Depropanizer/Debutanizer).

In commingle mode, the rich and lean gas will be mixed in the feed gas reception facility before entering LNG trains. The feed gas composition will be similar for all trains, hence the required modification will also be applied similarly for all trains. The LNG/LPG products from the LNG trains which have similar quality are sent to storage facility via rundown line. LNG/LPG transfer line and dock facilities are used for LNG/LPG loading purpose. Rundown line, storage tank, transfer line and loading docks are interchangeable.
Commingle hybrid is actually similar with commingle mode. But, in commingle hybrid mode, the LNG product from trains will be separated in different LNG storage tanks via separated LNG rundown lines. This condition is intended to accommodate LNG specification which required by a specific LNG buyer to maintain existing sales contract.

Some possible cases were developed to determine the impact of different feed gas profile to the LNG plant operation. The case consist of normal case and some sensitivity cases. The normal case is used as basis for production and plant thermal calculation, while the sensitivity cases are used for determining the worst condition of required plant modification.
The developed cases are as follows:

a. Commingle Normal
b. Gas Well A Trip
c. Gas Well B Trip
d. Gas Well C Trip
e. Gas Well D Trip
f. Gas Well E Trip
g. Gas Well F Trip
h. Gas Well G Trip

Simulation and Result
Simulation of LNG quality shows that some cases will experience the LNG density lower than 453 kg/m^3 starting 2018 which the lean gas start on production. The leaner gas into the plant will result in the lower LNG density.

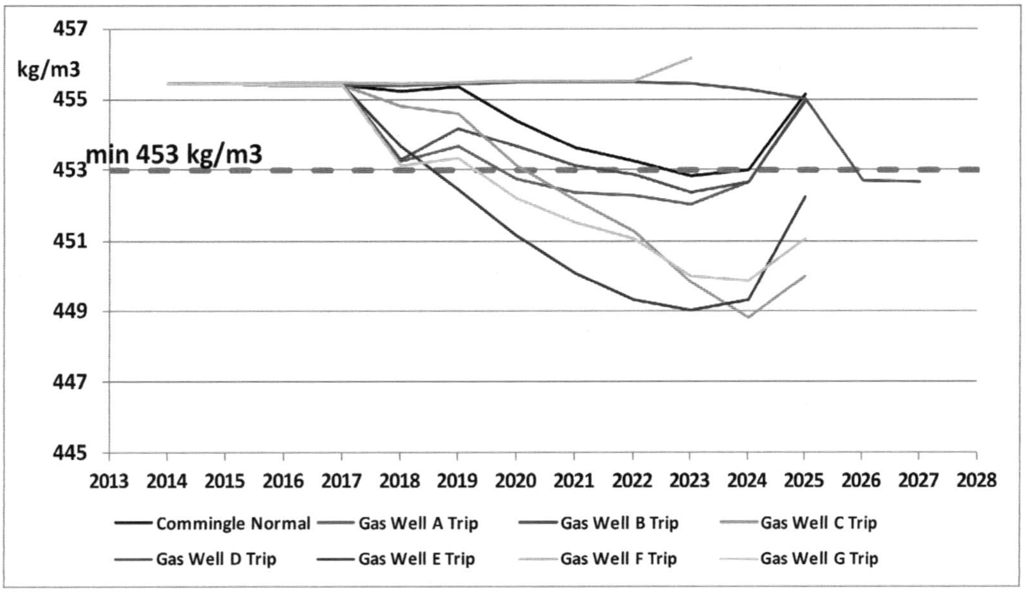

Figure 2. LNG Density

Similar with the impact on LNG density, the leaner gas in commingle mode will result in the lower LNG heating value. Most cases will also experience lower LNG HHV than 1,107 Btu/scf starting 2018, except the case of Gas Well F Trip. The profile can be seen in figure below.

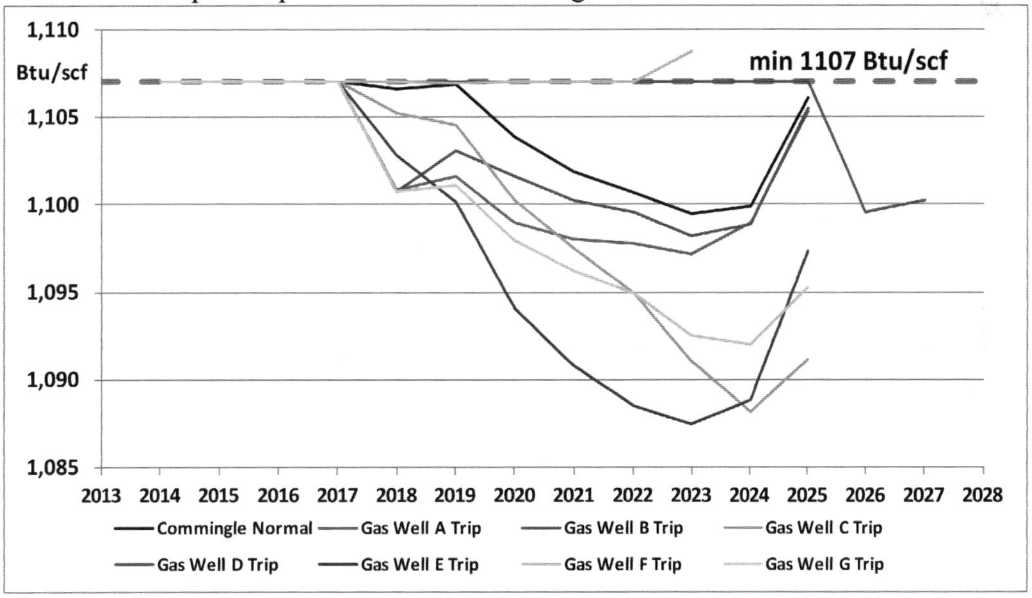

Figure 3. LNG Heating Value

HC freezing may occur due to high concentration of C_5+ and BTEX in feed gas inlet of Main Heat Exchanger. The Main Heat Exchanger manufacturer uses 3 parameters to identify the potential HC freezing by giving limitation of C5+, benzene and cyclohexane content in the feed gas inlet Main Heat Exchanger. The concentration is limited at 0.1 %/1.46 ppm/1.00 ppm for C_5+, benzene and cyclohexane, respectively. Once one of those components exceeds its maximum limit, HC freezing will potentially occur inside the equipment which service at cryogenic temperature up to -150 °C. Since all cases show that C_5+ and cyclohexane content are still far below its maximum limit compared to benzene, only benzene is used as parameter of HC freezing indication. The simulation revealed that most cases will experience heavy HC freezing, except the case of Gas Well D and Gas Well F Trip (all cases will have HC freezing in 2026-2027).

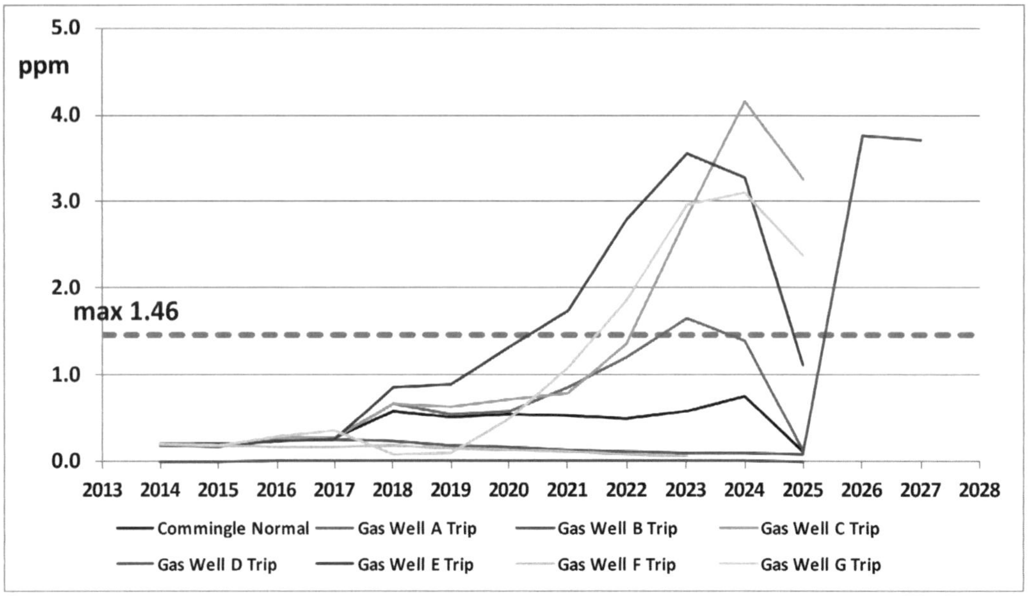

Figure 4. Potential HC Freezing

Ethane (C_2) refrigerant requirement for 1 train is 210 Nm^3/h. Based on simulation, C_2 production per train is still adequate to fulfill C_2 for refrigerant. Therefore, no issue for C_2 refrigerant requirement in commingle hybrid mode.

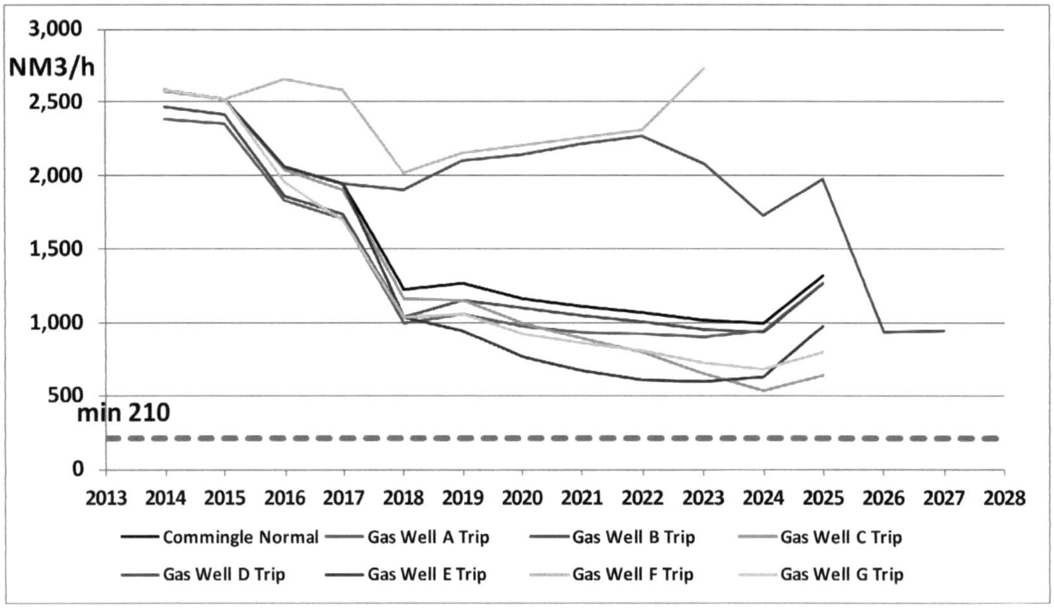

Figure 5. C_2 Production for Refrigerant

Similar with C_2 refrigerant case, no issue is found for C_3 refrigerant requirement for all cases. Propane (C_3) refrigerant requirement for 1 train is 381 Nm^3/h.

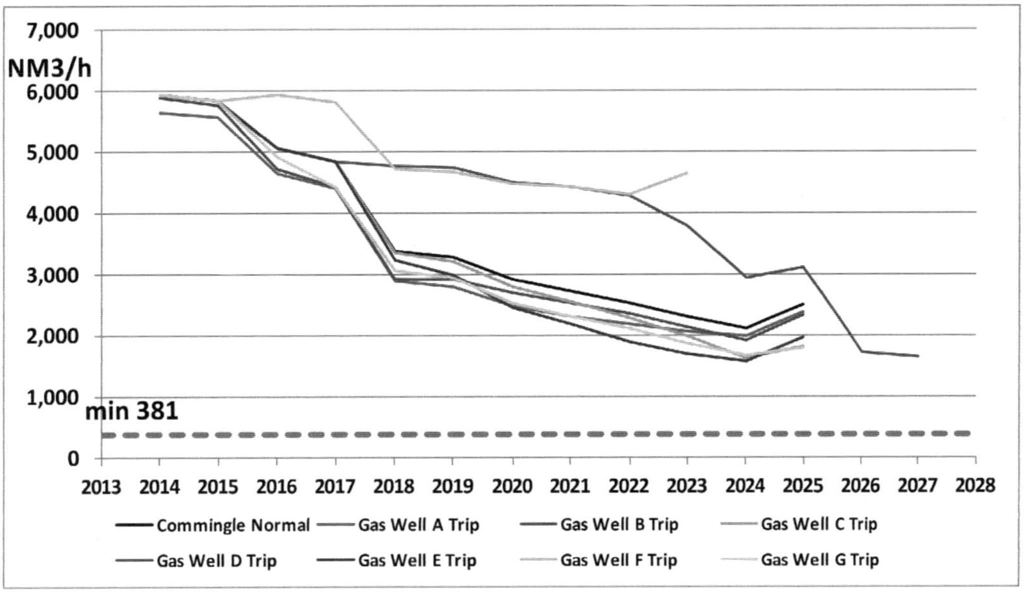

Figure 6. C_3 Production for Refrigerant

Fractionation columns (except Scrub Column) will run into lower turn down ratio due to incoming leaner feed gas. The leaner feed gas will impact to fractionation columns performance. The portion of gas as top product of Scrub Column will be higher. Since the vapor rate becomes higher, it potentially requires modification at the enriching section (feed to top) of Scrub Column. The internal tray will potentially to be replaced with the higher capacity type tray.

Besides causing higher vapor rate at the Scrub Column, leaner gas will reduce the portion of C_2+ as bottom product of Scrub Column hence the feed into the next fractionation column (Deethanizer/Depropanizer/Debutanizer/Splitter) will be lower and eventually reduces the load or column turn down. The column will experience turn down ratio reduction until 20%. Most columns limit the turndown ratio >40% to keep normal operation. The simulation shows that most cases will have fractionation turn down below 40%, except the case of Gas Well D and F Trip. All cases will also have turn down below 40% in 2026-2027.

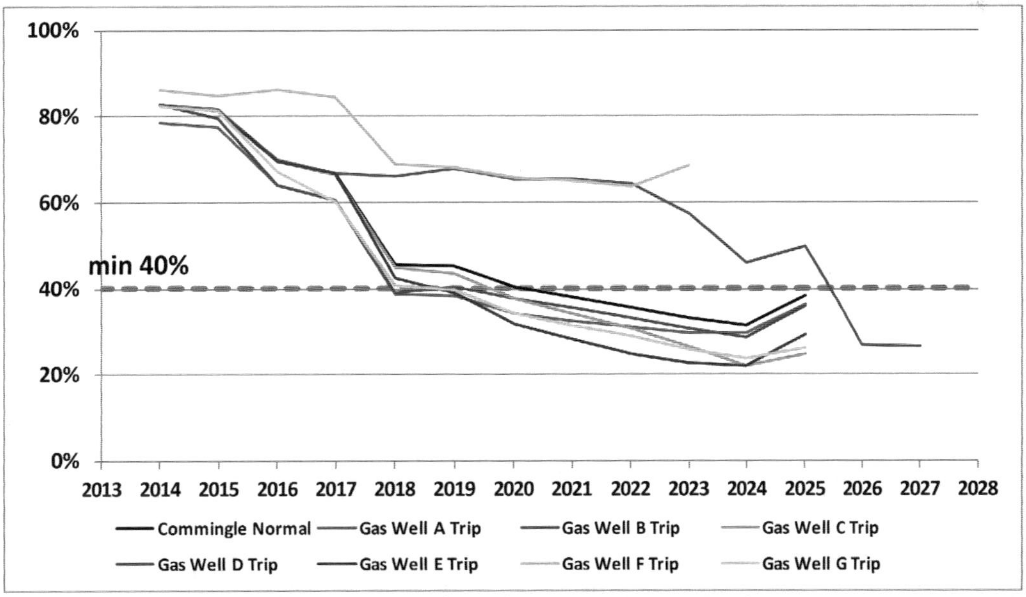

Figure 7. Fractionation Turn Down Ratio

Required Plant Modification

a. Additional Scrub Column Overhead Condenser

In general, the removal of benzene and C_5+ components to prevent freeze-out to meet the LNG product specification will be more difficult as the feed gas becomes leaner. Figure 4 shows the profile of benzene content at the inlet gas to Main Heat Exchanger. In current operation and feed gas composition, increasing the butane recycle to the Scrub Column helps to remove these components. However, the coldest Scrub Column condenser temperature achievable using low level propane refrigeration is ultimately limited. Once the butane recycle reaches 100% and this coldest temperature is reached, a plant modification will be necessary to remove the remaining benzene and C_5+ up to below its allowable limit. One potential solution would be to add a supplemental Scrub Column condenser using a slip stream of mixed refrigerant (MCR) to achieve a colder Scrub Column condenser temperature than propane could provide at above atmospheric pressure.

Additional condenser may give better operational flexibility. Cooling media for this condenser will use MCR liquid stream at temperature range of -115 ~ -130 °C. Based on simulation, temperature of -38 °C at feed gas outlet the condenser has been able to remove benzene and heavy HC to be lower than its maximum limit. However, safety margin should be added to give more satisfactory result. Meanwhile, this condenser outlet temperature should also be controlled not lower than -45 °C due to material concern at the downstream of the condenser. Temperature control should be installed to maintain the temperature higher than -45 °C.

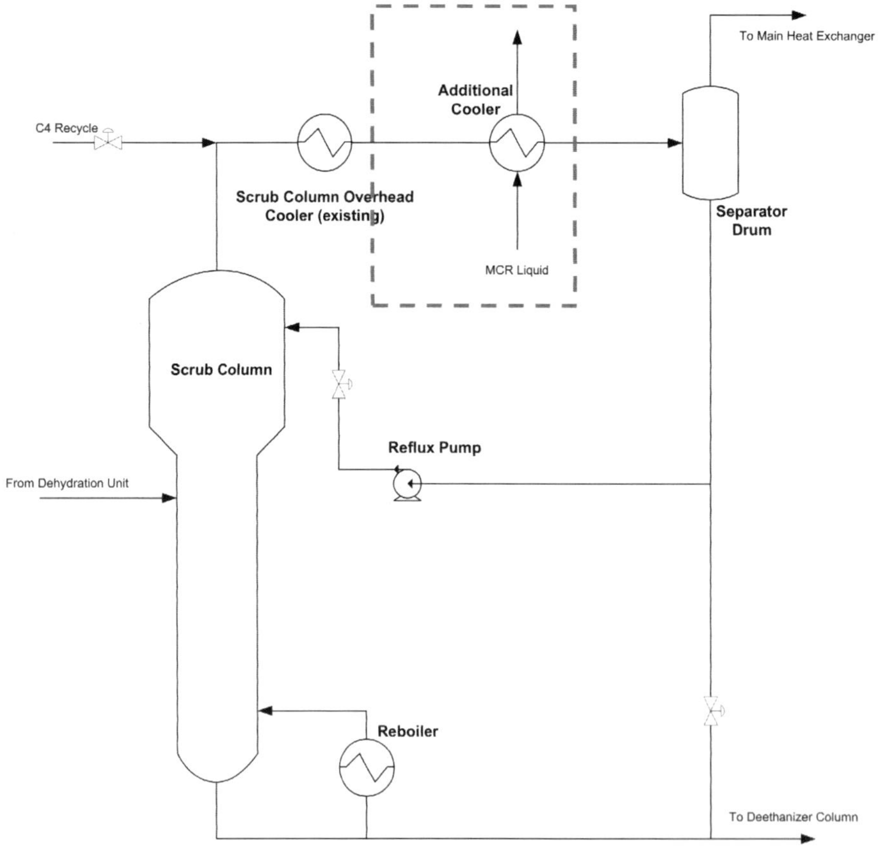

Figure 8. Additional Scrub Column Overhead Cooler

b. Tray Modification in Fractionation Columns

The leaner feed gas will impact to fractionation columns performance. The portion of gas as top product of Scrub Column will be higher. Since the vapor rate becomes higher, it potentially requires

modification at the enriching (feed to top) section of Scrub Column. The internal tray will potentially to be replaced with the higher capacity type tray.
Besides causing higher vapor rate at the Scrub Column, leaner gas will reduce the portion of C_2+ as bottom product of Scrub Column hence the feed into the next fractionation column (Deethanizer/Depropanizer/Debutanizer/Splitter) will be lower and eventually reduces the load or column turn down. The column will experience turn down ratio reduction until 20%. Most columns limit the turndown ratio >40% to keep normal operation.
Discussion with tray manufacturer was done to assess the possible modification in fractionation columns to overcome the lower column turn down ratio issue. The column should be operated in certain way with lower turn down without sacrificing the product purity. One of the options is by lowering the fractionation turndown ratio. It is possible to be done by internal tray modification (i.e : tray plugging). The manufacturer proposed to modify the Scrub Column top section for higher vapor rate (i.e : tray type replacement, replacing tray with packing).

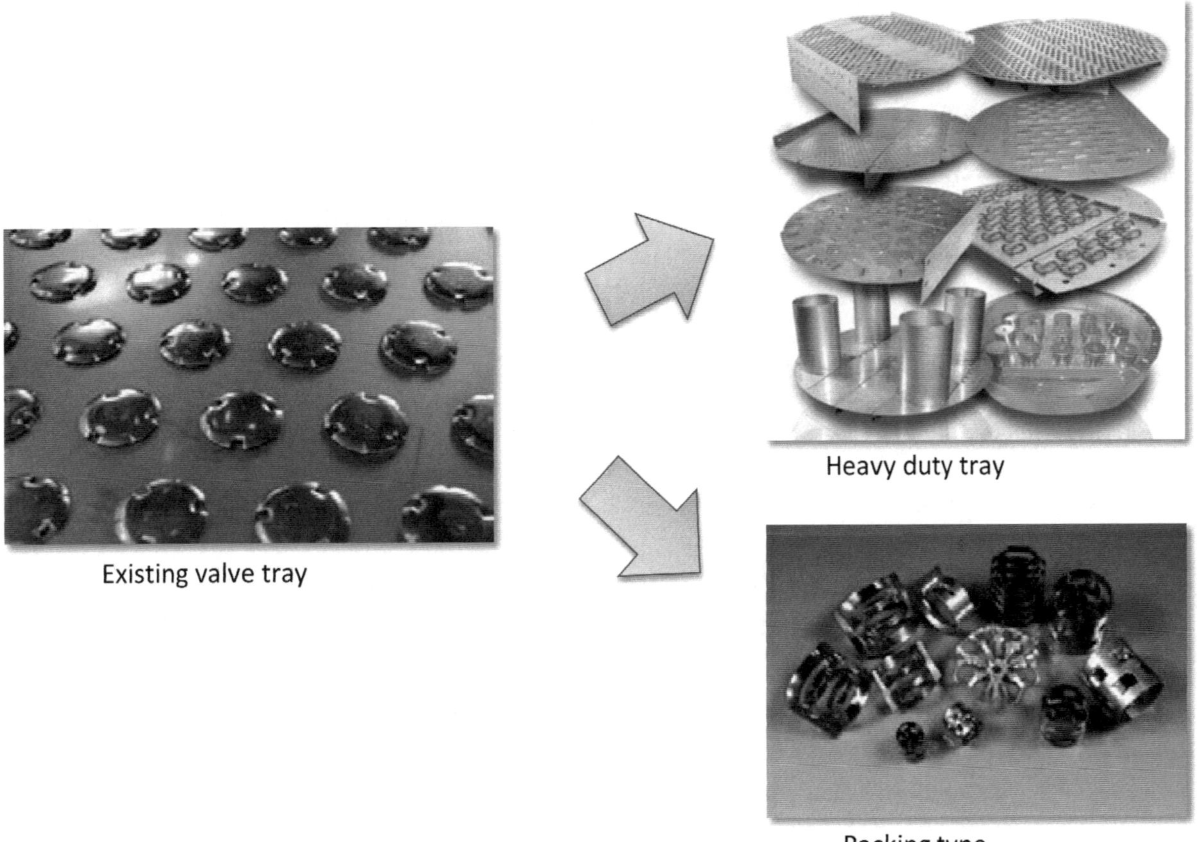

Figure 9. Possible Tray Modification at Top Section of Scrub Column

Tray replacement in Scrub Column is aimed for enabling the column to handle high vapor rate. The proposed solution from tray manufacturer is replacement by heavy duty tray or packing in top segment of the Scrub Column.

.

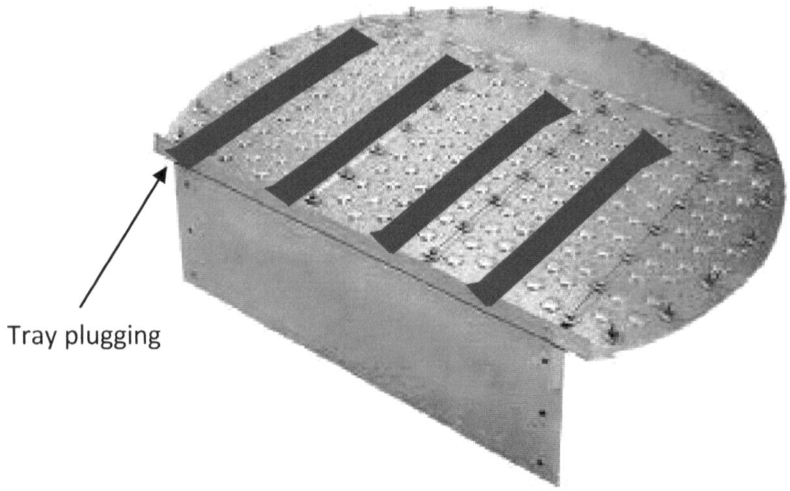

Figure 10. Possible Tray Modification at Deethanizer/Depropanizer/Debutanizer

Meanwhile, modification of fractionation column operating range (shifting capacity) needs further evaluation by tray manufacturers. Tray plugging is aimed for reducing the active area to allow a lower column turndown. In this case, existing tray will be maintained and some area of tray will be plugged to reduce the active area. As a result, maximum turn down capacity will be lower. Column shifting capacity by partial tray plugging is applied for Deethanizer, Depropanizer, and Debutanizer

Pros and Cons
Table below shows the advantage and disadvantage comparing previous mode of segregation, commingle original and commingle hybrid.

Table 2. Pros and Cons of Segregation, Commingle Original and Commingle Hybrid

Variable	Segregation	Commingle Original	Commingle Hybrid
LNG quality	Meet	Not meet	Meet
LPG Production	Higher	Lower	Higher
Plant Efficiency	Lower	Same	
LNG Tank Flexibility	Lower	Higher	Lower
Plant Safety	Same		
Plant Modification Cost	Lower	Higher	Lower
Plant Operating Cost	Higher	Same	
Production Revenue	Higher	Lower	Higher

Conclusion

1. Required plant modifications are fractionation column internal adjustment and installation of additional cooler. This modification is required to avoid column turn down ratio lower than 40% and HC freezing in the Main Heat Exchanger.
2. Commingle hybrid will give better result than other mode in term of LNG contract, additional LPG production, additional production revenue and less plant modification cost.

424667 Shale Gas - It's Commercial Viability and Rising Importance As the 'Next' Fuel

Abstract:

Shale, an ordinary semi permeable sedimentary rock has been no short of a miracle drug to the US energy sector. Shale gas is today regarded as the 'next' fuel for the entire world after the demonstration in the US, regarding its cleanliness in combustion and lower rates than conventional crude oil. The present work primarily focuses on the basic properties of Shale gas that builds it possible to be rehearsed and how it differs from other conventional sources. It also discusses about how the shale gas helps in reducing the effects of conventional oil and the other advantages that it has offered like reduction in CO_2 emissions. The work also focuses on the various techniques which are involved in the origin of the shale gas such as hydraulic fracturing and horizontal drilling. In the final, there is a discussion on the IEA approved practices and norms which should be strictly adhered to for sustainable extraction and usage of shale gas. The work derives that there is a requirement of new and stringent regulations to preclude deprivation of an historic opportunity to provide cheaper and cleaner energy to satisfy the world consumption, as easily as to usher in the future development of the shale gas industry.

407824 Experimental Study of Combined Use of Fe2O3 Nanoparticles and Magnetic Fields on Improved Oil Recovery

Abstract:

Experimental Study of Combined Use of Fe2O3 Nanoparticles and Magnetic Fields on Improved Oil Recovery

Silva, Roberto*; Molina, Daniel; Chaves, Árlex.

School of Chemical Engineering, Industrial University of Santander. Calle 9 and Carrera 27. P.O. Box 680002. Bucaramanga, Santander.

*Presenting author: roberto.silva.zea@gmail.com

The study intend to search new ways to enhance oil recovery from fields of Colombia and neighboring countries of the Andean Basin. A sandpack porous medium saturated with paraffinic oil will be used to simulate a sandstone reservoir type and the oil will be move throw out by the combined application of a nanofluids water base compound dispersed Fe_2O_3 nanoparticles at various concentrations and magnetic fields. The proposed methodology first will characterize the porous medium, fluids and the nanoparticles (NMR, XRD, etc.), then the porous medium oil with water and then with nanofluids and magnetic fields will move. The effluent collected will be analyzed for their composition and viscosity study after combined treatment and then study the mathematical relationship between the effect of treatment times, nanoparticle concentration and intensity of magnetic field applied. The validation will be based on results provided by Craig (1982) from secondary oil recovery by immiscible displacement and compare with water and surfactant displacement in similar porous mediums. It will be expected that the results show that exists a relationship between the increase in enhanced oil recovery from this proposed technique. The conclusions shall show the effect of the time of magnetic field applied magnetic field intensity and concentration of nanoparticles on oil viscosity to enhance its sweep and improve the recovery thereof compared to a conventional water displacement process.

Keywords: nanoparticles, magnetic field, enhanced recovery.

426878 Kinetic Analysis Using Master Plot Method and Distributed Activation Energy Model for Non-Isothermal Pyrolysis of Heavy Oil

Abstract:

Global energy consumption has increased steadily and it leads to continuously increasing production of crude oils, major source of energy. However, the production of conventional oils becomes insufficient to satisfy growing demand for oil due to decreasing recoverable reserves. Eventually, conventional oils are likely to be exhausted in the near future. Therefore, it is expected that unconventional heavy oils such as oil sands bitumen, extra heavy oil, and heavy residue would play an important role to meet oil demand, because they have potential to be converted into more valuable light oils. As the production of unconventional heavy oils is increasing, upgrading technologies become increasingly crucial for its effective utilization. Pyrolysis process, which is one of promising upgrading techniques, can be efficiently used to produce useful light oils from heavy oils. Kinetic analysis is important for the development of pyrolysis process because pyrolysis behavior can be predicted through obtained kinetic parameters, and kinetic parameters are fundamental for modeling pyrolysis process and designing reactors. In this study, pyrolysis behavior of heavy oil was observed under non-isothermal condition with different heating rates using a thermogravimetric analyzer in order to understand pyrolysis characteristics of heavy oils. Also, a proper reaction model was determined based on the master plot method. Kinetic parameters (activation energy, pre-exponential factor, and standard deviation) of heavy oil pyrolysis were subsequently determined using a distributed activation energy model (DAEM). Using DAEM, complex pyrolysis reactions of heavy oils can be described with continuous activation energy distribution. In addition, the activation energies are represented by a distribution function of activation energy, $f(E_a)$, for heavy oil pyrolysis.

430881 Structural and Molecular Characterization of Oil Sands and Heavy Bitumens Derived from Domestic Reserves

Abstract:

Non-conventional oils have become more attractive to meet energy requirement of the world since conventional oils constitute approximately only one-third of the total reserves and almost half of it already produced. Increasing oil consumption triggered by the growing energy demand results in a sharp decrease in conventional oils and opens up new alternatives for non-conventional ones, but they bring up new challenges in upstream, midstream and downstream processing as well.

Since researches try to overcome these challenges by developing new technologies, detailed characterization of these unconventional resources play an important role for their feasibility and processibility in the state-of-the-art upgrading technologies such as delayed cokers. These analyses are of paramount importance for the design of new processes as well not only for the conversion of heavier hydrocarbons into lighter and more valuable ones in downstream phase but also for the upstream and midstream phases.

In this study, non-conventional oils such as domestic bitumen and oil sands are investigated by means of elemental analysis, molecular weight analysis / Gel Permeation Chromotograpy (GPC), Proton Nuclear Magnetic Resonance (H-NMR), Differential Scanning Calorimetry (DSC), Thermal Gravimetric Analysis (TGA), Fourier Transform Infrared Spectroscopy (FTIR), X-Ray Diffraction (XRD), GC-Simulated distillation, American Petroleum Institute (API) gravity and Conradson carbon test. By means of the solvent sequence protocol, subgroups of these heavier feedstocks such as saturates, aromatics, resins and asphaltene content are also measured that is directly related with the compatibility and stability of possible feedstocks to be processed in refining operations.

The results indicate that these non-conventional oils can be used in current upgrading technologies such as delayed coking process but proper blends are to be prepared to meet the feedstock's processability limitations such as asphaltene and sulfur content. The physical and chemical analysis of these possible feedstocks will also provide crucial information for the prediction of the final properties of downstream products and also for the design of new upstream, midstream and downstream processes.

AUTHOR INDEX

AUTHOR INDEX